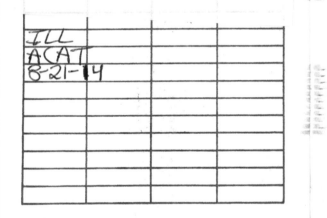

NATURE'S NIGHT LIFE

NATURE'S NIGHT LIFE

ROBERT BURTON

Blandford Press
Poole   Dorset

First published in the U.K. 1982 by
Blandford Press, Link House,
West Street, Poole,
Dorset, BH15 1LL

Copyright © 1982 Blandford Books Ltd

British Library Cataloguing in Publication Data

Burton, Robert
 Nature's night life.
 1. Nocturnal animals
 I. Title
 591   QL755

ISBN 0 7137 1111 6

Typeset by Polyglot Pte. Ltd.
Printed and bound in Singapore
by Toppan Printing Co. (S) Pte. Ltd.

Half title page: Long-eared bat,
Plecotus auritus, hovering
Title page: Long-eared owl, *Asio
otus,* eating mouse

Contents

The World by Night

'The dead of night' sums up the human view of the hours of darkness. Light is life, darkness is death. For us, the day is the time of activity; the night is the time for sleep. During the day we can see what is happening around us and can go about our business, but the night is a blank period, unless the darkness is banished by artificial illumination. We are so used to producing light at the flick of a switch that it is now difficult to appreciate how much the lives of our forebears were governed by the 24-hour cycle of light and dark. Only when a power cut throws everything into confusion is it brought home to us that mankind is truly diurnal.

This bias towards the day lies deep in our evolutionary history. The first primates were small, keen-nosed animals which scurried along branches at night. Eventually changes in the limbs and other parts of the body led to the appearance of the 'higher primates'—the monkeys, apes and man. During the transition, the nose was displaced as the principal sense organ by the eyes and the growing importance of vision over other senses became crucial. The eyes became more sensitive with the development of colour vision, and the eye sockets moved forward to give good three-dimensional vision. The higher primates were now very dependent on their eyesight, especially for safety and for finding their way about. With only one exception, the night monkey of South America, we and our simian relatives are strictly animals of the day. However, as will become clear, there is no simple link anywhere in the animal world between good vision and being active by day.

For us, the division of day and night is an obvious classification of the world but this is not so for all animals. Few are strictly diurnal or nocturnal. Nevertheless, although the night is largely outside our own experience, the fact that so many animals are active at night shows they deserve to be studied for their own interest and as an investigation into a general biological problem: why should some animals be more active at night and others by day? There are several reasons for avoiding the daylight. They are not always obvious or clearcut and it may not be easy to decide what factors are determining an individual animal's pattern of behaviour. This aspect of natural history has not often been studied in any detail.

Our diurnal nature has certainly coloured our appreciation of the world around us. Unless a conscious effort is made, it is easy to forget that the country does not retire to rest when we go home in the evening and awake with us the next morning. In the grey light before dawn, the country does seem to be asleep.

A night's hunting for this dog fox has produced a rabbit for the cubs.

7

The end of the night is heralded by a loom of light on the eastern horizon. This turns pink and suffuses the sky until there is a pallid glow against which birds appear as dark silhouettes flying out from their roosts. Then the first twitterings and snatches of song are heard and, as the light grows to illuminate the lower levels, grazing animals appear, first as flat silhouettes then becoming solid and substantial as the light improves.

This impression of the world awakening is, however, misleading. The birds, the most conspicuous of animals, are diurnal like us but the mammals have been out for much of the night. A vigil at dusk helps to rectify this impression of a sleeping world. It makes us realise that there is, if anything, more activity at night than by day. Even the birds, which seem at first glance to be an outstanding exception, do not sleep solidly throughout the hours of darkness. As we start our vigil the light fades first to a glow, then to a suffused glimmer. Birds stream into their roosts and the world becomes seemingly empty of movement. Once darkness has fallen it is almost impossible to see what is going on during one of the busiest times for animals. There may be quiet scufflings as small creatures push through fallen leaves, while bats flit overhead, and night animals call to one another but there are few other signs of the bustle and stir that is at least the equal of what was seen during the day. Half the year is night and the naturalist who regularly goes home at sunset is cheating himself of so much interest.

A little thought reveals the variety and wealth of animal life which can be encountered by night. Starting indoors, the animals which share our homes usually come out when we have gone to bed and the house is in darkness. By and large, they are not welcomed. The house mouse, a native of the Mediterranean but now a worldwide companion of man, makes its presence known from scattered droppings, nibble marks on food, the familiar 'mousey' smell and

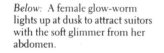

footprints in dusty corners. It is active by day but is more active at night and sharp scrabblings behind skirtings can be attributed to its movements.

Probably more unpopular, and for less reason, are the spiders which scurry, eight-legged and hairy, with equal ease across floors and walls but are often trapped by the slippery sides of the bath. They spend the day in silk-lined crevices, waiting for unwary creatures to wander onto their fine silk cobwebs. Before hygiene and cleanliness had such an important place in the home, cobwebs were valued as an instant dressing for minor cuts (they are quite clean when fresh) and the 17th-century philosopher Robert Burton approved of wearing an amulet containing a spider for protection against fever.

The house spiders' prey includes silverfish and firebrats. These are bristle-tails—small wingless insects, probably very like the animals which gave rise to

10

House mice can be a pest when they eat and soil food, or nibble holes in the skirtings. They come out to feed when the household is in bed, unless food is scarce and they are not disturbed.

Below: The silverfish is another pest in the larder. It is a primitive insect which scurries out of sight when the light is switched on.

the entire insect group many million years ago. The silverfish has a covering of shiny scales which give it a metallic appearance. It is found all over the world, outdoors in the tropics but confined to buildings in temperate regions. Damp places such as bathrooms are preferred and it feeds on fungi and decaying scraps. The firebrat has certainly suffered from increased hygiene. It needs the considerable warmth, up to 40°C, of kitchens and bakeries and these places are now much cleaner than formerly.

Hygiene has similarly hit cockroaches and house crickets. Originally an inhabitant of North Africa and Asia, the house cricket is another animal which has been spread around the world. Although usually innocuous and even considered by some to enliven the evenings with its chirpings, in recent years hygiene has largely driven it out of doors. It is no longer 'the cricket on the hearth' as the poet Milton described it, and it has taken refuge in piles of vegetable rubbish where the heat of decomposition helps to keep it warm.

The cockroaches, whose name is derived from the Spanish *cucaracha*, are serious pests despite sustained efforts to get rid of them. They, too, are warm-climate insects which have found human habitation to their liking and have spread around the world. The two best-known types are the 1 cm German cockroach and the giant 3–5 cm American cockroach. They infest warm, damp, dark basements and kitchens and come out at night to eat, and ruin, food, paper and clothing. They are often a problem in restaurant and hotel kitchens where they can be seen running for shelter the instant a light is switched on.

The interiors of the best kept houses may be virtually free of small lodgers but a walk outside, even in an urban garden, usually reveals traces of one or more nocturnally active animals. Hedgehogs forage on the lawn, bats fly overhead, foxes and owls call from nearby, and everywhere in the garden the light of a torch reveals a host of lesser animals—insects, woodlice, slugs and snails. Moths flitter around on their flowery business for most of the year but they are most abundant on mild nights. Attracted to the light, they gather especially at lighted windows and if observed closely through the glass from indoors, their eyes shine as pinpricks of light. This eyeglow, or eyeshine, as it is called, is a common feature of nocturnal animals, from cats and bushbabies to elephants and crocodiles.

The very fact that these animals come out at night has been a barrier to studying them, yet the mystery of their routine attracts our attention. The traditional time for the naturalist to watch nocturnal animals is at the time of their emergence, while there is still some lingering daylight and a watch kept at the resting place or a regular pathway gives a good chance of finding the animals.

Badger-watching has almost reached the level of a national pastime in Britain because badger setts with their gaping entrances, large soil tips and well-trodden trails make badgers easy to find and observe. The setts are often dug in sandy or chalky soil so that the badgers' movements can be discerned against a pale background and their habit of playing and grooming before departing on their nightly forays gives the observer the opportunity of a rewarding encounter.

Continuous watches at setts over the year have shown that the emergence of badgers more or less coincides with nightfall. The appearance of the badgers is delayed by a couple of hours after dusk in winter and is usually before dusk in the summer but the timing is also influenced from day to day by whether there is

Above: A lover of warm places, the house cricket chirps through the evening by running a row of pegs on the right-hand front wing over the rear edge of the left-hand front wing.

A dog fox brings a rabbit to his vixen who is confined to the den with a family of young cubs. Like many carnivores, foxes hunt by night or day, but human disturbance often restricts their activities to the hours of darkness.

Badgers leave the sett at night. In dim light their striped faces show up when the rest of the body is invisible. The stripes may be for identification or to warn other animals that a badger is a stern adversary.

bright moonlight or an overcast sky, by the availability of food and by certain aspects of the social behaviour of the animals themselves. The return to the sett just before sunrise shows a similar variation through the year, with the badgers retiring well before dawn in winter and sometimes returning in sunlight in summer.

It may be mentioned in passing that the obvious drawback to studying nocturnal animals is the 'unsocial hours' which the work demands. I must confess that my enthusiasm for badger-watching, a solitary pastime, used to wane as spring advanced and the later sunset and emergence of the badgers curtailed my own social activities. A more insidious drawback is the slowing of the body metabolism at night. Vitality is at its lowest in the early hours. This is the time that old people die, and even healthy field workers lose their ability to concentrate and react quickly.

Much of our ignorance of the night life of animals is now being dispelled through the growing availability of technical aids for tracking and observing. When an animal cannot be followed directly, its movements can be tracked by equipping it with a small radio transmitter. The researcher follows it with a portable receiver and a directional aerial which gives the orientation of the animal. Basically, radio-tracking allows the position and movements of individual animals to be traced. A bat, for instance, can be tracked from roost to

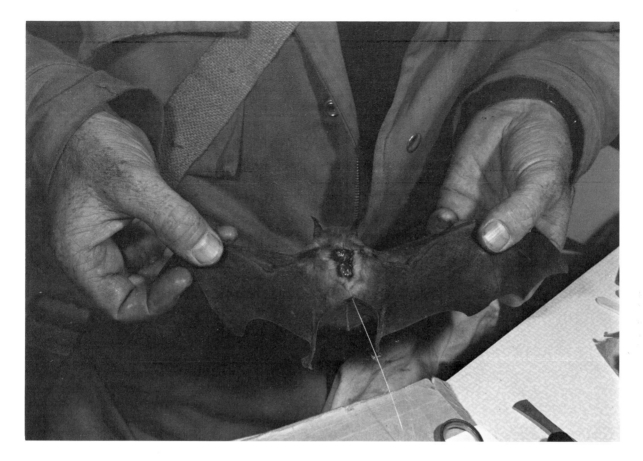

Where does a bat go when it leaves its roost? A tiny radio transmitter glued to its back signals its movements. Eventually, the radio falls off and is recovered.

roost, and from roost to feeding ground and back. Without the aid of the radio transmitter, it would have been almost impossible to have found out that in Britain male hedgehogs roam over an area of some 20 hectares, more than twice the range of the females, and that they cover a total distance of about 2.5 kilometres in a single night. Householders, who put out food for them, may talk about 'their' hedgehogs; but they are shared by the neighbourhood!

Radio-tracking can be made more sophisticated if the transmitter broadcasts more than the position of the animal. A thermistor in the circuit will, for example, alter the transmitted signal in response to temperature changes. When the transmitter is secured under the neck of a squirrel or a rat—like the brandy barrel on a St. Bernard dog—it gives different signals if the animal is curled up warmly in its nest or is running about with the radio hanging free.

Once an animal is located, its nocturnal activity can be kept under observation with night-viewing devices which allow a human observer to watch an animal at night without disturbing it. The first attempts to do this were simple and exploited the fact that some nocturnal animals are insensitive to red light. A lamp with a red filter was all that was needed. The method works well for watching mice and rats coming to feed on baits of grain but the range is short and does not work for all animals. More success has been achieved with the use of the

An infra-red lamp throws light on the behaviour of nocturnal animals. Special binoculars turn infra-red light into visible light.

infra-red 'sniperscope' developed in World War II. This uses a lamp which emits infra-red light instead of the white light of a normal electric lightbulb. The illuminated scene is observed through an infra-red sensitive telescope or binoculars, and the impression is of watching a black and white film.

A newer device is the image intensifier or 'starlight scope', a device which magnifies the natural illumination of moon and stars, or the glow from distant lights. The scene is bathed in an eerie green light, the brightness of which depends on the amount of natural light. The starlight scope's range can be increased by fitting tiny lights to the animals under study, but this has its drawbacks: the lights are likely to effect the animals' behaviour and they must be a positive beacon for predators!

Some researchers still prefer the older infra-red method. Because it shines a beam of light, the animals will cast a shadow which makes them easier to pick out when scanning the scene with the special binoculars or telescope and some animals become very conspicuous because their eyes glow as readily in the infra-red light as in car headlights.

A sea otter opens a clam by smashing it against a rock balanced on its chest. Until night-viewing equipment was used, the sea otter was not known to feed at night.

An example of the value of night-viewing devices comes from the Pacific coasts of North America where the sea otter dwells. It spends almost its entire life at sea, diving for shellfish and riding out storms in tangles of floating kelp. It had always been believed that sea otters were diurnal, although there had been a few reports of their diving by night. Night-viewing equipment has revealed that they have regular bouts of feeding throughout both day and night.

As a result of these modern techniques a picture is now beginning to emerge of night life, the unseen half of nature, and this serves to emphasise that the division of animal activity into the two categories of diurnal and nocturnal is an oversimplification. The cat is held up as an example, pure and simple, of a nocturnal animal but cats often hunt in broad daylight and even such renowned nocturnal animals as badgers and bats occasionally emerge into the sun. Clearly, animals do not always fit into a simple pattern of being active either through the day or through the night. Yet it is also apparent that some animals are principally active at night even though they can be seen at other times. We may therefore surmise that patterns of activity are varied and that there are advantages to such arrangements. We may also surmise that animals which are active by night are adapted to cope with the problems of nocturnal activity.

These may appear at first sight to be obvious conclusions, yet they form the background to a situation that is complicated in the extreme.

Adapting to the Dark

It has long been known that the lives of all living organisms, from the very simplest unicellular organisms, as well as plants, and including man, are organised on a rhythmic basis. In many instances, this rhythmic behaviour is clearly related to regular changes in the environment. For example, breeding times of most animals are tied to the annual cycle of climatic change, so that offspring are produced at the time of year when their survival is favoured by optimum weather conditions and food supplies. The monthly lunar cycle and the related rhythms of high and low tide are shorter cycles whose influence is felt most by organisms living in the tidal zone of the coast, but a lunar cycle also appears in a number of unlikely places such as the responsiveness to light in certain freshwater fishes and the human menstrual cycle.

Superimposed on any of these cycles of life is the 24-hour rhythm which seems to be one of the fundamental mechanisms of the natural world. The alternation of light and dark is the most obvious regular variation in the environment and the 24-hour rhythm is the pattern of activity which underlies the subject of this book. Over two centuries ago, it was discovered that organisms would continue to show a 24-hour rhythm even when isolated from changes in light and temperature. The experiment was originally carried out on plants which 'go to sleep' by lowering their leaves. The daily cycle of furling and unfurling continued unaltered in an unchanging environment. Since that time similar experiments have been carried out on a wide range of organisms. The most spectacular, in the form of an endurance record, was performed by human volunteers who descended caves for months at a time. With no clues about the passage of time from changes in light, temperature, humidity or any other perceptible environmental element, they continued to show rhythms of sleep and body temperature fluctuation. Sometimes the rhythms became stretched but they usually remained remarkably close to the 24-hour cycle.

Such feats of endurance are unnecessary to demonstrate the persistence of daily rhythms. Whenever a shift-worker changes to the night shift, the natural daily pattern of environmental temperature (warm in the day, cool at night) is reversed in relation to the cycle of work and sleep. Yet the worker's regime of body temperature fluctuations persist. Both cave-dwellers and the shift-workers must, therefore, have some internal basis for their daily physiological rhythms. This is the 'biological clock', a timing mechanism within the body which regulates the internal rhythms of body temperature, digestion and hormone secretion, and the observable patterns of sleep, feeding, drinking and other behaviour. The exact nature of the biological clock is not known. It is not a specific organ which can be dissected out or tampered with, but it appears to be a

18

mechanism diffused through the body tissues and has been found in every major group of living things, with the exception of bacteria.

The '24-hour clock' which controls the daily cycle (there must be other clocks for different rhythms) does not run entirely true. An animal kept under constant conditions in the laboratory deviates slightly from an exact 24-hour rhythm. For this reason, it is usual to refer to daily rhythms as 'circadian', meaning 'about a day'. The biological clock's 'day' is slightly more or less than the 24 hours of a mechanical clock so that cycles of activity gradually shift in relation to true time.

This shifting does not happen under natural conditions because the biological clock is continually reset by environmental rhythms, particularly by the transition from light to dark or vice-versa. By changing the time of the transition in the laboratory, the setting of the clock can be altered and the animal's active period pushed one way or the other. Cockroaches, for instance, become active at about nightfall and reach a peak of activity early in the night. A simple experiment is to keep them in a box illuminated by a light bulb. If the light is switched off at 6 p.m. each evening, after continuous illumination all day, the cockroaches reach their peak activity a few hours later. Then, if 'lights out' is changed to 6.30 p.m., the peak is shifted to half an hour later within a few days of the new regime being inaugurated. The animals can, however, only cope with small shifts in time: changing 'lights out' by several hours results in chaos.

The ability to alter activity in response to environmental changes ensures that the biological clock is geared to events in the outside world. If the cockroaches relied entirely on their internal clocks, even if they kept to an accurate 24-hour period, their lives would soon become deranged. They have to take into account the seasonal change in daylength so that they do not become active in broad daylight during the summer months.

The timing of activity periods within the daily cycle can be studied conveniently in the laboratory. At its simplest, caged animals can be watched and their movements noted. This is very time consuming and various methods have been devised to record activity. The ease with which small mammals take to exercising in tread wheels can be put to good use. A small dynamo, attached to the wheel and connected to a suitable recorder, gives an accurate picture of the exact times that the animal has been active and how fast it ran during its active periods.

Showing that an animal kept within the confines of a laboratory cage maintains a certain pattern of activity demonstrates that the pattern must be important to the animal for it to be so deeply ingrained in its behaviour. Yet the sudden snapping on and off of an electric light is unnatural and does not mimic even the fastest tropical sunrise or sunset. A more accurate simulation of natural events is gradually to lower and raise the light intensity over the period of an hour thereby imitating dusk and dawn. When this scheme was first tried out on a variety of mice, the changes in behaviour proved to be quite dramatic. As the light started to fade, the mice suddenly began to run at speed, slowing down as the light dimmed. At artificial dawn, the mice ran faster as the light grew brighter, then abruptly stopped and retired to their nests. In between, when light levels were maintained at the equivalent of moonlight, the mice continued at a low level of activity.

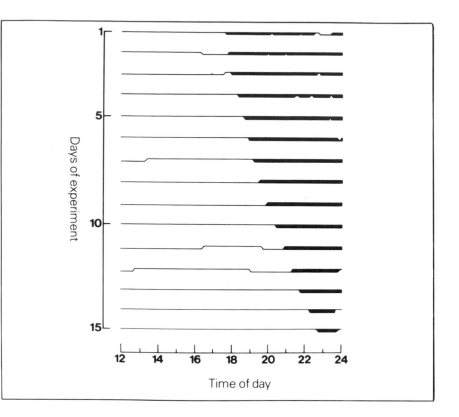

The activity rhythm of a flying squirrel is 24 hours 20 minutes long. When one is kept in darkness for 15 days, it becomes active 20 minutes later each night, as indicated by the black bars (*After* Bünning, *The Physiological Clock*, Springer-Verlag).

In other words, these mice were basically nocturnal but they were most active at dawn and dusk. Such behaviour is termed crepuscular and further experiments have shown that several carnivores also exhibit similar patterns of activity. It is clear that crepuscular behaviour, with two peaks of activity during the twilight hours, is a common phenomenon and that nocturnal behaviour is not a simple switching on and off of activity any more than sunset is equivalent to switching off a light. In some instances, activity consists of a main peak when the animal fulfils most of its daily obligations, followed by a smaller peak after a rest period. The latter may last for less than an hour and be very variable in timing and extent. Even the main peak is variable in that it may commence at a set time, such as sunset, but continue for a longer or shorter period on successive nights.

The simple notion of nocturnal, diurnal or crepuscular activity is further complicated by the habit of many animals to be active for short bursts of activity throughout the 24 hours. This can be seen in small rodents, carnivores and insectivores and in hoofed mammals which have alternating periods of feeding and resting. This habit is adopted either because, like the hoofed animals, their digestive systems require the continuous passage of large quantities of low-grade food, or because, like small flesh-eaters, they need frequent meals to support their high metabolic requirements. Nevertheless, there may be a 24-hour cycle imposed upon the short bursts, so that there is more activity, overall, by day or by night.

Whatever may be the pattern of an animal's activity, the ultimate aim of the zoologist is to find out why the animal should have adopted such a pattern. This is done by comparing its activity with changes in the physical environment and the animal's relations with food, predators and members of its own kind. In practice, this has not always proved practicable and there are many instances where an animal's behaviour has remained something of a puzzle, but there are two general points which can be made. The persistence of the patterns in laboratory conditions, when the environmental clues of food, predators and so on have been removed, demonstrates that they are very deep-seated. The rhythms must, presumably, have been fixed in the animal's genetic make-up and have been moulded by the pressures of natural selection so that the animal is behaving instinctively to the best advantage. On the other hand, the pattern is often flexible and the animal can adapt rapidly to changes in its environment. Carnivorous animals, in particular, do not maintain rigid regimes but arrange their hunting to coincide with the time that prey is available. This may entail a shift in activity when they switch from one quarry to another but the most common reason for behavioural change is associated with man's activities.

As a diurnal predator or mere disturber of the peace, man has driven many animals to become nocturnal. When lions were being shot in the Kruger National Park in a misguided policy of predator removal, they responded by becoming more nocturnal in their habits. Presumably they had learned that hunters did not operate at night. The same reaction has been observed in elephants and other animals which have suffered from persecution. In a survey of

Whiskers and nose assist the otter's eyes when it is fishing at night or in murky water.

Large, sensitive eyes are useful at close quarters, but feathery antennae guide male moths to a female hundreds of metres away.

night life this poses the problem of discerning which animals are truly nocturnal and which have been forced into a retiring lifestyle by human intervention. The European otter is, according to the books, nocturnal, yet I have watched otters in broad daylight from the shores of remote Hebridean islands. They were fishing or basking on rocks with little concern at my presence and I am not alone in having had the opportunity of watching these delightful animals which have a reputation for being elusive. On the peaceful coasts of the Scottish Islands, where otter hunting was never commonplace, they have suffered little persecution. Elsewhere in the country, active hunting and general disturbance by riparian activities have turned the otter into the shy and all too rare animal portrayed in *The Wind in the Willows*. The true nature of the otter's pattern of behaviour could be revealed by laboratory experiment, using a tread wheel. The method has been employed on animals as large as the wolf, and would give insight into the otter's natural lifestyle.

Exploring the nature of biological rhythms is a wide-ranging subject which goes beyond the study of nocturnal behaviour. It is sufficient here to show that different rhythms exist in a 24-hour context and that animals have several options for organising their activities. Nocturnal activity is just one possibility which has

22

its peculiar advantages and disadvantages, and requires certain adaptations in the structure of the animal to become practicable.

The essential attainment for animals which are active by night, whether as a committed way of life or as a temporary expedient, is that they should be capable of carrying out the functions of life in dim light. There are two major options for nocturnal animals: they can improve their vision, or they can abandon it and use other senses. By and large, hunting animals have taken the first course because vision provides a guidance system for finely judged pouncing on prey. Their victims have, in general, opted for the second course. They usually eat plants or other bulk food which does not need a fine discrimination to gather. Except for animals living in caves or burrowing in the soil, vertebrate animals do not lose the power of vision completely, but nocturnal rodents probably do little more than distinguish light and dark.

During the twilight hours spanning sunset the light falls by an incredible factor of one million, and by the time darkness has fallen on a moonless night, the light has been reduced 100 times still further. Yet it is not pitch dark. To quote Jules Verne: 'On earth, even on the darkest night, light never abdicates its rights. It may be subtle and diffuse, but however little light there may be, the eye finally perceives it'.

In the same way that animals can be divided into the three categories of diurnal, nocturnal and crepuscular in habit (with the proviso that this is a simplification), so their eyes are adapted for working in bright, dim or varying light conditions. The human eye is an example of an eye belonging to a strictly diurnal animal and although it works best in bright light, it shows that even a diurnal eye is not completely useless at night. Despite the long daytime pedigree of man and his ape and monkey ancestors, the human eye functions better at night than we often credit it.

On leaving a lighted room or while sitting outside in the evening, our eyes undergo the process of 'dark adaptation'. We say that we become accustomed to the dark. The process takes at least half an hour to complete and proceeds in several stages. The most obvious is the dilation of the pupil, in which the iris retracts to let the maximum amount of light into the eye and onto the sensitive retina. Using the photographic analogy, this is equivalent to opening the aperture of a camera in dim light, as indicated by reducing the setting of the f number. This takes a second or two and, in fact, makes little difference to the sensitivity of the eye.

Changes taking place in the retina and its associated complex of nerve networks play a much larger part in the increased sensitivity of the dark-adapted eye. Part of this is due to changes in the visual pigment, rhodopsin, which reacts to the stimulus of light entering the retinal tissue and stimulates the firing of nerve impulses to the brain. Rhodopsin is similar to vitamin A and a diet deficient in the vitamin can lead to night blindness. Both are related to carotene, the substance that colours carrots, hence the wartime story of night fighter pilots eating carrots to see better in the dark. Rhodopsin is gradually bleached by bright light but it regenerates in the dark to make the retina more sensitive. A day spent in bright sunshine, sailing or sunbathing without dark glasses to shade the glare off the sea, can slow dark adaptation by several hours, but the main mechanism of dark adaptation is in the nerve networks which process the signals coming from

Above: The eye of a toad shows how the iris retracts and the pupil dilates to make the most of dim light.

Below: Kirk's dik-dik caught by the flash. A small antelope, living in thorn thickets, it feeds at dawn and dusk.

Right: The epitome of a nocturnal animal. A black cat surveys the twilight scene with its sensitive eyes dilated.

the retina's light sensitive cells. Details of the mechanism are still a mystery but it is responsible for an increase in sensitivity approaching a millionfold.

The eyes of nocturnal animals are further improved in two ways: they are designed to collect as much light as possible and the retinal sense cells are so arranged that they are stimulated by very low light levels. The first essential for collecting light is a big pupil and every photograph of a nocturnal animal shows its large round eyes. Hand in hand with the large pupil, go a large lens and a bulging cornea. The lens may be almost spherical and its increased power intensifies the brightness of the image falling on the retina.

When a nocturnal animal wishes to bask in the sun, as many do, the retina has to be protected by the pupil closing to a vertical slit, as in the familiar example of the cat. In very bright sunshine, the cat half closes its eyelids to cut down the light even more. Frogs, bony fishes and birds have a second protective mechanism: the retinal sense cells are partly covered by pigment in bright light which moves back, like a curtain drawing back, in dim light.

A good proportion of the light which enters the eye passes straight through the retina without stimulating the sense cells, but in nocturnal animals this lost light is intercepted and reflected back through the retina by the tapetum, a layer of tissue which acts as a mirror. The reflected light has a second chance to stimulate the sense cells so that the eye's sensitivity in dim light is greatly enhanced. However, this is achieved at the expense of acuity, the double stimulation resulting in a blurred image which, although better than no image at all at night, would be an unnecessary hindrance in daylight. Truly diurnal animals never have a tapetum.

Even on the return passage through the retina, some light still escapes and shines back through the lens as 'eyeglow' or 'eyeshine'. This is the well-known phenomenon of the eyes of cats, foxes and other animals glowing at night when caught in headlights. Eyeglow is usually green or yellow, but is sometimes red and, with experience, animals can be recognised by the colour, height above ground and pattern of movement of glowing eyes. On the African savannah, pairs of bounding red dots belong to springhares; cats' eyes in general are golden yellow and lions' are large; antelopes' are large and white; and hippos' are small, red and set wide apart.

The retina contains two kinds of sense cells: rods and cones. Both contain substances which absorb light and trigger nervous activity. Without going into the details of their functioning, the rods are used in the perception of dim light, while cones are used for discerning fine detail in bright light and are also involved in colour vision. Part of the rods' sensitivity in dim light is due to the way in which they are 'wired' into the network of nerve fibres in the retina. Whereas each cone is connected to its own nerve fibre, a group of rods shares a common fibre. The result is that an image falling on the retina is analysed in detail by cones but poorly by rods. On the other hand, light which is too weak to stimulate a cone to trigger a nerve impulse, may be sufficient to persuade several rods to fire their common nerve fibre. Sensitivity of nocturnal eyes is increased by the linking of hundreds or even thousands of rods to a common nerve fibre.

The retinas of nocturnal animals are packed with rods at the expense of the cones which would be of little or no use in dim light. Bats, bushbabies and

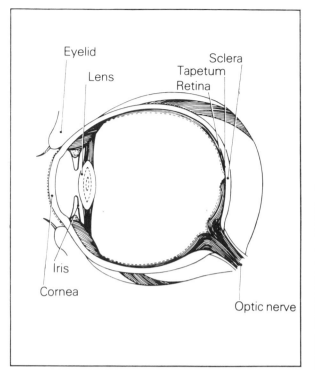

The vertebrate eye adapts for night vision by the iris retracting. This dilates the pupil and lets more light into the eye. Nocturnal animals have a tapetum to reflect light back through the retina. (*After* Marshall and Hughes, *The Physiology of Mammals and other Vertebrates*, Cambridge University Press).

Above right: A hare at night reflects the photographic flash from its widely dilated eyes.

nocturnal snakes and lizards have no cones, while some strictly nocturnal rodents have only a very few. Animals which may be active by day as well as by night have a retina of both rods and cones. Cats, which are known to possess colour vision, and human beings have mixed retinas. Our retinas have cones packed in the centre where the image falls from an object on which our gaze is concentrated. Rods predominate in the periphery of the retina and we can see things better at night out of the corner of our eyes. In practice it is difficult to see something when you are deliberately not looking at it!

There are, therefore, two basic ways in which the vertebrate eye can adapt for nocturnal vision. There are the changes which have resulted in the structural differences between the eyes of nocturnal and diurnal animals, such as size of the eye, aperture of the iris and proportion of rods to cones. Then there are the flexible changes which an eye can undergo as its owner passes from light to dark. These include the opening of the iris and adaptations in the retina.

The insects, which rival the vertebrates in supremacy for the Earth's surface in terms of numbers if not bulk, have evolved a system of vision which is very different from the vertebrate eyeball. Despite this, nocturnal insects have taken the same two options as the nocturnal vertebrates: some have relegated vision to a secondary role; others have developed extra-sensitive eyes.

The insect eye is described as compound; it is made up of a number of units or facets, each of which is an independently functional structure like a miniature eye, complete with cornea, lens and light-sensitive cells, and the underlying nervous system integrates information from different facets to create a coherent

27

The pupil's response to light is immediate. The pupil on the sunward side of this caiman has closed more than the other.

picture of the world. The number of facets is an indication of an insect's visual powers. Fast-flying, predatory dragonflies, which can distinguish movements over 10 metres away, have 28,000 facets packed into each eye. The worker ant crawling through the leaf litter, and finding its way by scent or feel, has only a handful.

Those insects which use their eyes at night have a variation of the compound eye called the superposition eye. The advantages of the superposition eye for nocturnal insects lie in the huge aperture, with an f number of 0.5 to 1.0, (compared with f = 2.4 for the diurnal honeybee) and the summation effect of the lenses. Instead of each lens having its own small group of sense cells isolated from neighbouring facets by columns of pigment, as in the ordinary compound eye, light can travel between the facets so the array of lenses forms a single image on a carpet of sense cells. As a single sense cell receives light from many lenses, as many as 100 in some moths, weak light has a greater chance of stimulating a response in the retina, in a manner recalling the combined action of rods in a vertebrate eye.

28

The pupil of the flying gecko has scalloped edges. When fully closed, light can only enter through small gaps where the edges do not meet.

While it is one thing to show that an animal has eyes designed for use at night, it is another matter to determine the role of eyesight in its life's activities. Even during the day, eyesight is of little use for finding food or watching out for enemies if the animal is living among tall grass or dense undergrowth. Yet the opposite is also true. Eyesight can be valuable even when it is too dark to see clearly. Try walking through a wood on a night which is not too dark. Instead of straining the eyes to look for obstacles, take in the general view. The outline of the tree canopy against the sky and the intense darkness of thick bushes are a good indication of the lie of the land and give the chance of unimpeded progress.

For a nocturnal animal, with its tapetum, wide-aperture iris and ultra-sensitive retina, this must be a reliable form of navigation through the countryside but it may not be enough for locating prey. Ernest Neal has described a serval prowling in long grass and using its ears to locate prey. The serval is a medium-sized cat, the African equivalent of the lynxes of Europe and America, with a beautiful spotted coat and long legs. This serval was advancing

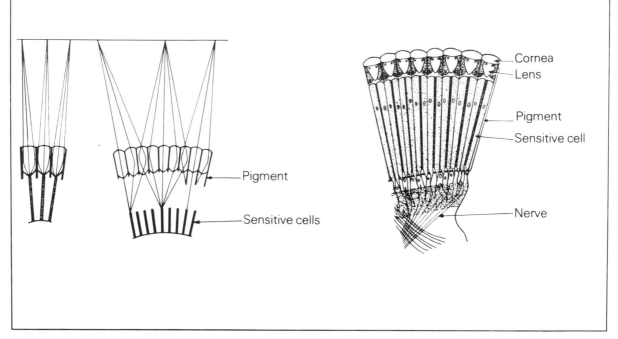

In a diurnal insect each facet receives light from a narrow field. Nocturnal insects have a superposition eye (right) in which light from several lenses is focussed onto one set of sensitive cells. (*After* Wigglesworth *The Principles of Insect Physiology*, Methuen).

On the far right is a section through the compound eye of an insect. It is made up of units or facets separated by pigment. Each unit has a group of sense cells grouped under a lens and attached to a nerve.

stealthily, placing each paw on the ground with great caution, while its head turned from side to side on the lookout for prey. Dr Neal could see its ears turning independently to pick up the slightest sounds and it soon detected the presence of a hare which must have been out of sight. The serval stalked forwards now with its ears coordinated to fix the precise position of its quarry. Finally, there was a 3 metre leap which landed the serval just short of the hare. Unfortunately for a neat story, the hare escaped, but all predators expect a proportion of misses, whatever senses are being used, and it must not be forgotten that the hares' equally sensitive ears will have been on the alert for sounds spelling danger.

Here, in the serval, is a cat with the usual nocturnally sensitive eyes of its family employing another sense for hunting. The same seems to be true for owls: despite their large, sensitive eyes, they appear to rely more on sound for locating prey. Compared with vision, it is difficult to show that other senses are adapted for use at night. The nose needs no modification to work in the dark, although it is reasonable to suppose that it may be more sensitive in nocturnal animals which rely more upon its use. Despite the nose's extreme sensitivity for discriminating ultra-low concentrations of an odour, smell is a coarse sense in that it has little locational capability for detecting the source of odours, except at very short range. A radical improvement of a sense as a substitute for vision has been achieved in a few instances. The bats and two species of bird have developed echolocation; certain snakes have an organ which, in effect, 'sees' heat; and several kinds of fishes use a sensitivity to electric currents for finding their way about.

30

A serval's ears turn back to show that it is alert even when licking its fur. Sensitive ears help to pinpoint prey in the dark.

Echolocation, which has also been developed independently by the dolphins for use underwater, represents a breakthrough in the use of the ears. However sensitive the ears of an owl may be, it can only locate prey which is making a noise, by squeaking or rustling through dead leaves, and it must still rely on vision to detect obstacles in its path. The bats, however, generate a beam of sound and listen for the faint echoes returning from prey and obstacles. The echolocation system is so sophisticated that bats can chase mobile prey at speed and virtually dispense with eyesight, although they are not so blind as the popular adage suggests.

The pit-viper family of snakes, which includes the rattlesnakes, is named after two small pores on the snout which are, in effect, simple heat-sensitive 'eyes'. They can detect the heat given off by warm-blooded prey and guide the snake's strike.

Echolocating bats and birds and heat-seeking pit-vipers are described in later chapters but the strange case of the electric fishes needs to be elaborated here. The ability of certain fishes to cause a numbing pain when touched has been known for centuries although it was not appreciated that this was caused by hefty electric shocks. The electric catfish of tropical African lakes and rivers can deliver up to 650 volts to stun its prey. The electric ray and electric eel are similarly endowed but several kinds of fishes produce currents so weak that they can be detected only by instruments. Fishes with organs generating weak electric

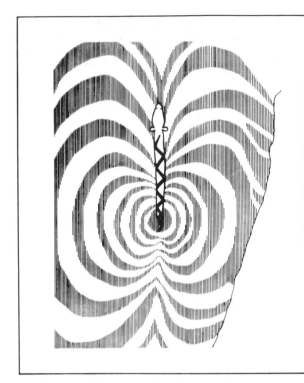

Right: The knifefish generates an electric field around its body to find its way in the dark. It swims with the long fin on its underside and keeps its body rigid so that the electric field is not distorted.

Above: Diagrammatic view of the lines of force generated by an electric fish. Objects in the field cause distortions which are detected by sense organs. (*After* Gray, in Marshall, *The Life of Fishes*, Weidenfeld and Nicolson).

currents include the elephant-trunk fishes and the Nile fish of Africa and the knifefishes of South America.

A common feature of these unrelated electric fishes is that they live in waters which are often turbid and they are mainly active by night. Their eyes are small and they are using their electricity to monitor their surroundings. The currents are generated in organs developed from muscles. The contraction of a normal muscle starts with a small charge over the surface of the muscle fibres, but in the electric organs of fishes, the power of contraction has been lost and the charges greatly increased. The fibres are no longer slender filaments but platelike and stacked like the cells in a battery. The function of the electric organ is to set up a field around the fish like that around a bar magnet, with a positive pole at the head and negative pole at the tail. Any object, animate or inanimate, in the water near the fish disturbs the field and the fish detects this by means of sense organs in the skin.

Whatever senses are used by an animal, whether it is diurnal or nocturnal, they have to be viewed in relation to the animal as a whole and its lifestyle. No animal will rely on a single sense, it will use an array of senses, each with its particular role. For instance, one sense may alert the animal to prey or predator while another guides its attack or evasion. The animal integrates and interprets information gathered by all its senses, adding the all-important memory of past experience, to build up a 'picture' of its surroundings and to act accordingly.

Animals are creatures of habit and rely on memory to regulate their lives. They spend a large part of their lives in one place, a territory or a range, which

32

they get to know intimately. They learn the safe resting places, the best sources of food and the whereabouts of neighbours. The advantages are obvious, for instance in terms of time saved in searching for food or shelter and in quickly slipping away from danger. If the animal knows its home patch well enough, it will be able to move about automatically. This is particularly important at night. Anyone who has come home late and crept up to bed in the dark for fear of disturbing the household, or has merely reached out for the bedside light, will know the value of familiarity with the surroundings. For blind people this knowledge is vital. Moving about the house during the day results in an unconscious learning of the distance between doors and furniture. It may not be completely accurate but landmarks such as a ticking clock or moonlight shining through a chink in the curtains give sufficient clues to adjust the course and avoid collisions.

If animals are better equipped than us to find their way about, it is because knowledge of their surroundings is so vital for their survival. The first act of an animal placed in novel surroundings is to explore. Whether it is a kitten in a new home or a mouse in a new cage, it moves slowly about its quarters investigating every feature. The accuracy of the learning process was demonstrated by a genet living in the Burton household. Genets are delicately proportioned members of the mongoose and civet family, and are found in Africa, in the Iberian peninsula and in southern France. They are among the most nocturnal of carnivores and they are arboreal. To patrol through the treetops in the dark requires extremely nimble footing and genets are assisted by extreme familiarity with pathways through the trees.

Our genet occupied a large wall-to-floor cage in one room. It was well furnished with boughs and perches and we could hear the genet padding around at full speed at night, when the room was in pitch darkness. How she learned the route was revealed when she was introduced to her quarters. She walked the length of every bough extremely slowly and with extreme caution. Each paw was lifted in turn and placed down. When its grip was secure, the next was moved

Portrait of a genet at rest. A nocturnal hunter, it moves through the trees using eyes, ears, nose, whiskers, a delicate sense of touch and a good memory of pathways.

forward. At the same time, all the genet's senses were alert. The large, delicate ears twitched this way and that, the nostrils quivered and her eyes stared in concentration as she peered around her. Having made one thorough examination of her surroundings, she repeated the circuit a little more quickly, following exactly the same route and still testing everything with her feet, ears, nose and eyes. On the third time round, the genet moved at speed and thereafter she showed complete familiarity with her new world.

The importance of memory in the genet's movements was demonstrated when she was allowed into an extension of the cage. This time she slipped at one point, lost her footing and hung momentarily from the branch before swinging herself up. Thereafter, she invariably somersaulted at this point as she ran along the branch. Reliance on memory has its drawbacks: novelty can lead to downfall. Bats can be caught in nets whose strands are thick enough to be detected by echolocation. It seems that the bats are so used to flying along a particular stretch that they do not notice echoes from the net and fly straight into it. It is not too hard to think of parallels in human life.

In showing something of the extent of nature's night life, including the way that animals organise their activities around the clock and get around the restrictions caused by the absence of sunlight, a very important question has been ignored so far. Why should animals choose to be nocturnal? According to biological teaching, there is a reason or function for all of an animal's attributes, so that some advantage for a nocturnal lifestyle should always be apparent.

While gathering material for this book I have often been frustrated by the lack of information on an animal's nocturnal behaviour. Sometimes this is simply due to the difficulty of observing secretive animals living in wild country. Yet even for well-known and familiar animals, the advantages of nocturnal lifestyles and the adaptations needed to maintain them have gone unstudied. The scientific literature is usually silent on the subject and when I have asked the opinion of experts about their chosen animals, the reply tends to be: 'Good question, but we do not know'. At most, the response is speculative discussion but not a clear answer, because in most cases the problem of nocturnalism has not been considered. Books I have consulted have, in their ignorance, described an animal bluntly as nocturnal, but people who know the animal well tell me that it is seen by day as well, or that it prefers the twilight of dusk and dawn, and the simple pattern of behaviour I am looking for does not exist.

As more people study more animals in greater detail, this book, like every other natural history book, will become obsolete. The study of animals under natural conditions and the manner in which they fit into their environment is proceeding apace. It has already led to an unevenness in this book. The animals of East Africa—lions, elephants, antelopes, zebras and so forth—are among the best studied in the world and the intimate details of their lives are presented in popular books and on television, but there is still plenty of mystery about what happens in the vastness of tropical forests and the depths of the sea, or even in our own backyards. The next chapter examines some of the reasons which animals have for adopting the nocturnal habit and it shows that the answer is by no means always simple and straightforward. It is an exciting subject because it offers unlimited scope for observation and speculation.

The Cover of Darkness

Sufficient examples have been given in the preceding chapter to show the extent of nature's night life and the ability of animals to move about and have their being in the dark. Plenty of animals are known to be active under the cover of darkness and laboratory experiments show how this pattern of activity may be ingrained in their behaviour. Yet the question why they should choose this way of life and what benefits they derive from it are by no means clear. The daily rhythms of a wide variety of animals, large and small, warm-blooded and cold-blooded, predators and prey, have been studied, but the descriptions and discussions of these in the scientific literature give no easy explanation why some animals are active by day, others by night.

Sometimes one comes across a nice story, like Guggisberg's account of how some young captive crocodiles became nocturnal when threatened by fish eagles, whilst others, whose enclosures were fitted with protective cages, remained diurnal. More often it is impossible to see such a correlation of cause and effect. This is largely because observations of the animal's life have rarely been directed at the actual study of nocturnal activity, the author being merely content to remark, in passing, that the animal is nocturnal. Analysis of its patterns of activity may show an inherent disposition for a nocturnal regime, its senses or other attributes may be assessed in terms of adaptation for operating in darkness, and the stimuli responsible for its actions may be discussed, but usually there is no indication why the animal adopted this way of life in the first place.

In some instances the reason may be obvious, as in desert animals which stay under cover to avoid the heat of the day and emerge only at night, but more often the suggestions, which are acceptable in principle, have never been investigated and remain as untested theories. The explanation of nocturnal behaviour may be made more difficult because there is the possibility of several underlying causes. This is, again, compounded by the lack of specific studies. In the case of the bats, for instance, it could be they are avoiding competition for the available food supply or avoiding predation but it may be simply that they cannot tolerate the bodily discomforts of hunting on sunny days. Flying is physiologically very demanding. The birds can cope with flight in warm air, and it may be that bats are at a physical disadvantage during the day. As far as I know this idea has never been raised.

In the following pages, the causes of a preference for nocturnal habits are examined in turn, with examples of their role in the lives of animals. From the start, we can propose three main reasons for adopting a night life:

1. The physical conditions at night can be more favourable for general activities. This is especially important for soft-bodied invertebrates and amphibians whose skin is not fully waterproof. The effects are most severe in hot, dry countries where even the best adapted animals find it expedient to avoid the heat of the day.

2. Darkness can give safety from predators, but the very existence of a source of food attracts predators which can hunt in the dark. In spite of this, the dangers of being caught are reduced, so the cover of darkness gives some immunity from attack.

3. The night shift can reduce competition for food. There is a general rule that animals tend to share the available resources by eating slightly different foods, by feeding in different places or at different times.

In this chapter we shall consider mainly the physical conditions of the night and their influence on the lower forms of animal life.

For many animals the night air is preferred because it is cooler and more humid. After the heat of the afternoon, the air temperature starts to drop and falls more rapidly after sunset. Thereafter there is a gradual fall until sunrise brings new warmth. The humidity of the air matches the changes in air temperature. The lower the temperature, the less moisture the air can hold and the greater the humidity. So as the evening temperature falls, humidity climbs. The magnitude of these changes depends on climatic conditions. When the sky is overcast, there will be little change in temperature or humidity over the 24 hours, whereas on a clear night, the heat absorbed during the day is radiated into the sky, the temperature falls considerably and dew may be deposited. This is particularly noticeable in deserts where the nights are surprisingly chilly, and the dew is important for the survival of desert animals.

A tropical millipede curls up when removed from its retreat. It is active at night when the air is cool and moist.

European eels find their way into lakes and ponds by slithering overland at night, when the ground is wet with dew.

Topi, large antelopes of the African savannahs, prefer pastures of long grass and, where there are few trees, they are exposed to the full rigours of the sun. They minimise the effects of the heat by feeding in the coolness of dawn and dusk and lying down during the day. They even avoid chewing the cud at this time because this doubles the body's production of heat. The other advantage of feeding at dawn and dusk is that the grass becomes moist with dew so the antelopes can make good the water lost during the rest of the day.

At the other end of the scale, cool, moist air is particularly vital for soft-bodied animals, such as slugs and snails, woodlice and the many kinds of worms. When the ancestors of these invertebrate animals left the seas and made the difficult transition to a life on land they faced the problem of drying up through the evaporation of their body fluids. An impermeable skin is the solution, but many of these animals have been only partially successful in achieving full water-proofing, as can be seen when a slug or worm stranded on a dry pavement soon withers and dies. Even the tougher-coated woodlice are not immune.

So, the danger of losing water restricts many soft-bodied animals to damp,

37

dark places. They live in the soil or in the layer of leaf litter above it, they creep under stones or under flaky bark and they emerge only when night has fallen. The common representatives of these animal groups are usually seen when their hiding places are overturned by the gardener and they scuttle or creep in search of safety. They can be watched in a more relaxed state by searching the garden with a red light on a warm, damp night.

Earthworms spend most of their lives underground where they literally eat their way through the soil. They come to the surface on occasion, stretching out over the soil but leaving the tail in the burrow to ensure a quick retreat. They are very sensitive to vibration, which could denote the presence of an enemy, and anything but the lightest footfall sends them back into their burrows as if pulled by springs. With a quiet approach it may be possible to avoid disturbing them and to hear faint rustlings as the worms probe around for dead leaves. These they pull back into their burrows to be nibbled, and sometimes daybreak finds the lawn marked out with leaves standing erect among the grass like miniature banners.

Snails can gain some degree of protection from drying up by retreating into their shells. In times of drought they seal the entrance with layers of mucus and so can survive for long periods, even for years, in a form of suspended animation. This has enabled some snails to colonise deserts, where they are hard to find until a shower of rain brings them out. However, they are still vulnerable to desiccation when they unfurl their bodies to creep about in search of food, so a nocturnal life is still a necessity. Slugs, which are no more than naked snails, are even more susceptible to losing water and need a secure daytime retreat in dry weather. Although a higher humidity is the vital ingredient of the night environment, it is the stimulus of falling temperatures which goads slugs and snails into activity. When rain or mist keeps the temperature low after daybreak, they stay out and continue feeding.

The inference is that the nocturnal habits of worms, snails and slugs are linked with the need to keep down water loss. They emerge when they will not be subjected to the drying effect of the sun's rays and when the humidity in the atmosphere is highest, helping to reduce the loss of water from their bodies. Some slugs are active by day. We would expect to find that they have better waterproofing than their nocturnal counterparts but this has yet to be determined. If such a correlation could be made, it would make a nice example of animals adapting to their environment, such as has been demonstrated in laboratory experiments on woodlice.

If a dozen or so woodlice are placed in a deep dish, with a glass plate to stop them climbing out, and a dark cover is placed over half, the woodlice move under it, thereby demonstrating an antipathy to light. Another experiment is to line the dish with soil, bone dry in one half and damp in the other. The woodlice wander about at first but they eventually come to rest in the damp half. These two simple experiments explain why woodlice are found under stones and loose bark. Woodlice also like to huddle against each other or their surroundings. They come to rest in corners of their dish, especially if they can get their backs under an overhang, and in natural conditions they gather in cracks and crannies.

Temperature and humidity are crucial to the survival of woodlice, but the response to light, shown in the first experiment, is what really controls their

Oniscus asellus is a woodlouse which dies quickly if its surroundings are dry, so it lives under logs or stones and comes out at night.

activity. Partly through an inherited circadian rhythm and partly through the triggering effect of nightfall, woodlice come out at night to forage for dead and decaying plants, and occasionally tender live seedlings. Movement starts some time after nightfall and continues through the night, until daybreak sends these crustacean Cinderellas running home.

The raised humidity and lowered temperatures of night reduce the loss of the woodlouse's body fluids. During the night the sensitivity to low humidity decreases and the woodlouse can wander into drier places, but the animal is then living on borrowed time with its body fluid evaporating away as inexorably as sand trickling through an hourglass. The time that can be spent away from shelter varies between species, however, depending on the degree of water-proofing.

There is gradation among the common woodlice of British gardens and woods in the degree to which they are nocturnal. For example, the very abundant *Oniscus asellus* is very susceptible to drying up and is strictly nocturnal. This species is a shiny brown-black with pale edges and is shield-shaped, half as broad as long. The pillbug, *Armadillium vulgare*, is readily identified by its habit of rolling into a ball and is more rectangular when uncurled. It is more resistant to drying up and is seen abroad in the morning. A third species, the common *Porcellio scaber*, is intermediate in its waterproofing. Even so, a Dutch biologist has shown that the relationship between nocturnal activity and the moist night air is not as simple as seems at first sight. He found that *P. scaber* is forced to move from its hideaway because of too much humidity. Woodlice absorb water from their surroundings as well as losing it by evaporation. Their excretory system cannot cope with the amount of moisture absorbed from saturated air, and they become waterlogged unless they move into the open to lose the excess by evaporation. On a humid night, then, woodlice may be more active not because they can spend more time abroad before drying up but because it takes longer to lose their excess water.

Even this unexpected state of affairs may not represent the full explanation of the woodlouse's behaviour and it illustrates the danger of inferring that nocturnal activity is the result of something obvious and simple such as avoiding drying up. A deeper investigation by experiment or observation may confirm the correlation, but ever more often modern biologists are finding that the natural world is turning out to be even more complicated than their predecessors had believed.

The two groups of invertebrate animals which have developed a complete waterproof covering are the insects and arachnids. The latter group comprises the eight-legged spiders, scorpions and harvestmen. Physical factors may still affect their lives, especially in deserts where there are extremes of high temperature and low humidity, but relations with other animals in the form of predation and competition are probably more important, except in the case of harvestmen.

These strange animals, which look like a bun slung from a frame of impossibly slender legs, are known as Daddy-long-legs in America, although this name is more commonly used for craneflies in Britain. They are found all over the world and most species are nocturnal. Compared with spiders and scorpions, harvestmen have inefficient waterproofing and frequently drink water to supplement the fluids sucked from their victims.

Spiders and scorpions, venomous animals which prey on small animals, are predominantly nocturnal. An exception is the wolf spiders which, using their huge, goggling eyes, seek and run down their prey by sight during the day. The many web-building spiders detect their prey by vibrations set up in the web and so are independent of light. The advantage of a mainly nocturnal activity is that these creatures do not so readily fall prey to larger predators such as birds.

Shrubs and hedges throughout Europe are decorated in the summer by the orb webs of the garden spider, also called the cross spider from the pattern of dots on its body. On dewy mornings the webs become spangled with droplets and the spiders show up as dark blobs in the centre, where they have been waiting all night. By thus taking up station in the middle of its web, the spider can immediately locate trapped insects from vibrations their struggles set up in the radially arranged strands of silk. As the sun rises and the dew evaporates, however, the spider retreats under the shelter of a leaf. It is still alert and holds a thread of silk like an angler holding his fishing line to feel for a bite. A

The early morning sun finds a
garden spider at the centre of its
dew-spangled web. As the
morning progresses, the spider
retreats out of sight.

struggling insect caught in the web brings the spider rushing back to the centre to deal with its prey. The insect is subdued, trussed up in silk and sucked dry of its body juices; but this is a dangerous place for the spider and it spends as little time as possible in this vulnerable position.

In considering the huge array of insect species, two million in all, there are definite problems in establishing why some should be nocturnal and others diurnal. Insects are the most highly developed group of invertebrate animals. They have penetrated all parts of the world and have taken up an almost unthinkable variety of habits, many of which are proving to be highly intricate and subtle, yet the simple matter of why they should be active at certain times of the day is usually overlooked. The night is full of insects: mosquitoes, moths, caddisflies, beetles, craneflies and others which can be seen gathering at lighted windows. Fireflies reveal their presence by their pinpoints of light, while bush crickets and cicadas fill the air with their trilling songs, but many more unassuming insect varieties will be quietly going about their duties without attracting attention.

Insects have proved to be the invertebrate group most capable of activity during the day. The impermeable wax-covered cuticle renders them virtually independent of atmospheric moisture and some are capable of surviving high temperatures. Yet many are nocturnal. Tropical ants, for example, are often nocturnal especially when they live in deserts but, even in deserts, there are some diurnal species. Insects, therefore, are not constrained so much to use the night as refuge from an adverse environment, except at certain moments in their lives when they are especially vulnerable. Thus, night is a favourite time for insects to moult their skins or to emerge from the pupa, leaving them with a new skin, that is soft and not fully waterproof. They are then not only in danger from desiccation but are also extremely vulnerable to predation and the night gives them protection.

Among the insects there are some groups which have both diurnal and nocturnal representatives. The Lepidoptera springs to mind first, with the basic distinction of butterflies flying by day and moths by night, although some moths are diurnal and the South African evening brown butterfly flies at dusk and dawn. The mosquitoes are another group with day- and night-flying species.

When we see different species of a group of animals spacing their activity over the 24 hours, we may suspect that there is no special advantage to day or night. Changes of temperature and humidity are not usually critical for insects in temperate latitudes and predation may not exert any great pressure one way or the other because there will be predators of one kind or another hunting at all times. An insect which retreats to the night to avoid birds will only have to face an onslaught from bats. As these animals have the sensory capabilities to go about their business as well in the dark as in the light, the division of light and dark is meaningless for them. What we are observing, therefore, is a division of the 24 hours between related species so that they do not compete with each other for food supplies.

Entomologists often make use of a light-trap to collect night-flying insects. The trap consists essentially of a powerful lamp to attract insects which then tumble, bewildered, down a surrounding funnel and into the trap. A very comprehensive study of night-fliers was made with a light-trap by C.B. Williams

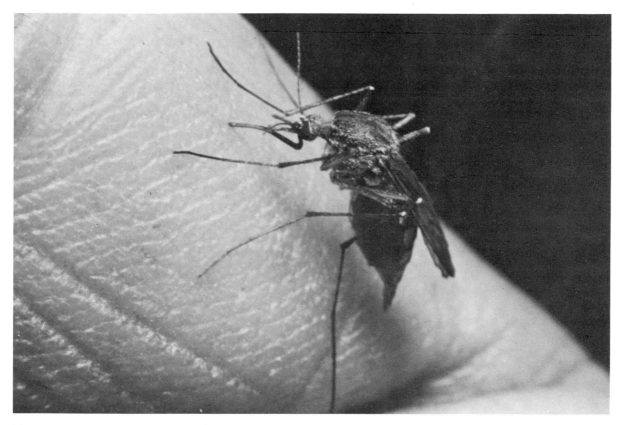

Mosquitoes are an ever-present menace in many parts of the world, as shifts of different species bite at different times of day and night.

Opposite: Termites live in the permanent darkness of their nests and tunnels and only swarming queens and kings emerge into the open. They choose the cover of the night for their single mating flight.

nearly 50 years ago. He ran his light-trap every night for two years, dividing each night into eight equal trapping periods: four before midnight and four after. This system meant that the length of the trapping period varied according to the time of year but that the first and last trapping periods always covered dawn and dusk, when the crepuscular insects are out.

The results of this thorough, time-consuming study showed that many species have definite periods of activity at a certain time of night. Some have the twin-peak pattern of activity at dawn and dusk; others appear during one period before midnight or before dawn. These patterns have been recorded for moths, mosquitoes, caddisflies and other insects. Well-defined, regularly occurring bouts of activity must, it would seem, have a precise function but this is very hard to establish.

The behaviour of mosquitoes has been studied in detail, especially in tropical countries, because of their importance as carriers of malaria and other serious diseases. Mosquitoes, as is well known, are in evidence at any time of day or night, but what is less well known is that different species bite at different times. The standard method of studying biting cycles is to employ human bait, who stoically allow the mosquitoes to land and pierce their skin while the researcher picks them off. Some species prove to be diurnal, some nocturnal and some are crepuscular, biting at dawn and dusk, so there is no respite from their painful activities.

The moult is a time of danger for insects. They are weak and helpless so the cover of night is helpful.

There is a similar distribution in time among mosquito species in their egg-laying and swarming. Swarms of mosquitoes and midges can often be seen, bouncing like tiny diaphanous yo-yos on invisible strings near some conspicuous object such as a tree, a building, a person or over a boundary of contrasting light and shade—a roadside verge for instance. The swarms are composed of males, and females come to find a mate. They are most common in the evening and in one species (*Psorophora confirmis*) at least, swarming is triggered by a fall in temperature to 11°C. Its swarms start within half an hour of sunset and continue for another half hour.

Swarms of midges and mosquitoes gather over landmarks. Some species swarm by night, others by day, but little is known about the function of the swarm.

Despite an impressive knowledge of the timing of mosquito activities, very little is known about the cause or function of what has been observed. Among crepuscular species the dusk peak is the greater, probably because the air is warmer than at dawn. Synchronisation of mating behaviour may be the function for swarming at a particular time but coordinated egg-laying and biting is less easy to explain. It has been suggested that *Aedes africanus* bites after sunset when its monkey hosts are settling in their tree nests and that species of *Mansonia* bite at night when birds are roosting. However, it is quite possible that this is the wrong way round. These mosquitoes may not select a biting time to coincide with the availability of their hosts, they may attack this host merely because it happens to be available at a predetermined biting time. Also there is no general rule of mosquitoes setting a mealtime when their hosts are most vulnerable. On the other hand, the dangerous, disease-bearing mosquitoes are true man-biters. They steal silently into houses at night to bite the sleeping occupants and their peak of biting is in the early hours when their hosts are least likely to retaliate. The troublesome, whining mosquitoes which advertise their presence are less well adapted to attack humans because they bite when their prospective hosts are alert and quick to swat them.

The invertebrates probably invaded the land under the cover of darkness to avoid the desiccating effects of the sun, and a similar course appears to have been followed by the first land vertebrates. Amphibians, which still start their lives as aquatic animals, are very susceptible to drying up. The skin is not waterproof and the surface is kept moist by secretions from glands which are dotted over the body. Water is lost from the skins of some amphibians as rapidly as from a free water surface.

The moist skin is essential to amphibians because their lungs are inefficient and much of the exchange of oxygen and carbon dioxide takes place through the skin. This means that the amphibian has to draw on its body fluids to maintain the continually evaporating layer of moisture and the result is that amphibians are more abundant in damp, tropical forests where a high humidity reduces water loss. Elsewhere, they are usually found near water, although some live in deserts where they survive by burrowing into the cool, moist, deep layers of soil. Water passes through the skin in both directions so that the frog or toad can make good any loss merely by sitting in water or wet vegetation, rather than by drinking. This was proved to the satisfaction of scientists only a few years ago although Dr Townson of London had come to the same conclusion in 1799.

In deserts and temperate regions the appearance of frogs and toads after rain, to the extent that the ground appears seething with them, is a common phenomenon. At other times, they are rarely seen because they are lying up under stones, in crevices and clumps of grass or other refuges, but they become active by night. The edible frog, for example, spends most of its time in water, but it sometimes comes out to hunt in the dark.

There is one time of year especially when amphibians make themselves heard. In temperate countries this is the spring when they return to water to court and spawn. Where frogs and toads are still common, and this is a shrinking area as development and industrialisation destroy lakes and swamps, they can be located by the chorusing calls coming from pools and puddles.

The European common toad migrates to its spawning pools at night and it is usually seen only where the route crosses a road. At such spots, there is a steady but very slow stream of toads crawling across the asphalt. Their numbers are so great that it is impossible to avoid running them over and in recent years animal lovers have started rescue operations to help the toads across the road.

The migration takes place at night, when the atmospheric conditions are most favourable and the toads are safe from predatory birds, although not from nocturnal predators such as otters. By day all that is to be seen is a scattering of flattened, dried bodies of road casualties. The males lead the procession and

The tree frog blows out its throat when calling. Choruses of frogs fill the air after nightfall in the mating season.

gather in the pools to await the larger females but there is some overlap as the first females may arrive at the water carrying a male, pick-a-back, in the mating position.

Nightfall is the signal, all over the world, for the gathered frogs and toads to start calling and they can be nerve-rackingly monotonous. The American bullfrog is named after its deep, booming call which can be heard over distances of $\frac{1}{2}$ kilometre. The bullfrogs spend the day hiding safely along the banks and come out in the evening to start calling. Apart from their calls, they advertise their presence to the females by inflating their lungs and floating high in the water to show off the yellow chin. They would be just as conspicuous to water birds, if these had not gone to roost by this time. In East Africa, where a variety of frogs and toads gather in newly formed pools at the start of the rains, the splashing of the amphibians and the noise of the tinkling and croaking calls attracts storks, egrets and herons. After the initial excitement, the frogs and toads seem to settle down and confine their courtship to the night and so escape the further attention of the birds.

One of the few investigations into the reasons for the nocturnal habit of

Above: Frogs breathe through the skin which must be kept moist for this purpose. Nocturnal activity reduces the amount of water lost through the skin.

Right: The tuatara of New Zealand prefers the cool temperatures of night, so it hunts at the time when petrels are visiting their nest burrows.

Opposite: These West Indian anole lizards are bright green or yellow during the day. By night they change to a subdued colour and sleep among the foliage.

amphibians supports the anti-predator theory. The American toad *Bufo americanus* likes warm, damp nights but it is less active when the moon is full. Darker nights confer some protection from accidental encounters with skunks and raccoons, although this will not save the toad from being smelt out.

Two of the advances which reptiles made when they evolved from amphibians were the full waterproofing of the skin and the provision of a parchment covering for the eggs so that these could be laid on land. Effectively, the reptiles became independent of water except for the need to drink. They were able to colonise dry places and no longer had to take refuge in moist places and restrict most of their activity to the night. Indeed, the reverse is true, for reptiles are sun-seekers.

Like the amphibians but in contrast to birds and mammals, the reptiles are cold-blooded. This means that they lack the ability to maintain a high, steady body temperature independent of their surroundings, as compared with ourselves who, for instance, maintain an even 38°C. This could be a misleading simplification. The real difference is that birds and mammals have a rapid metabolism and generate heat in the body which is conserved by a layer of fat under the skin supplemented by a clothing of feathers or fur. Reptiles and other cold-blooded animals metabolise slowly and cannot retain their body heat. Nevertheless, some reptiles maintain a fairly even temperature, higher than their surroundings, by their behaviour. This is most obvious in desert reptiles which bask in the morning and afternoon to soak up the sun's heat, shelter in the shade at midday and burrow into the sand at night to keep warm. A more accurate description of reptiles than cold-blooded is 'ectothermic' meaning 'using outside heat', while the birds and mammals can be called 'endothermic' or 'using inside heat'.

On account of this dependence on environmental heat, it is not surprising that reptiles, amphibians and other ectothermic animals should be most abundant in warm, sunny countries, and it is to be expected that they would be diurnal. That many are nocturnal shows, however, that the heating effect of the sun is not vital. In open country, warmth is rapidly radiated from the ground at night and ectothermic animals cool off. As the temperature drops they gradually slow down and, eventually, stop altogether. This is less likely to happen where a cover of thick vegetation keeps the ground and the lower levels of the atmosphere warm and ectothermic animals can keep going. Ultimately, of course, the heat comes from the sun, so such animals are much more dependent on daylight activity in cooler climates.

Soaking up heat from the ground or air is not the only way in which some reptiles manage to remain active at night. Many are known to have a 'preferred temperature' at which the metabolic processes of the body work most efficiently. Preferred temperatures vary between species, presumably because of physiological differences, and this is reflected in their patterns of activity. Many diurnal lizards, for instance, have preferred temperatures between 33 and 37°C, whereas nocturnal lizards prefer 25 to 32°C. (It should be pointed out that although these temperatures are preferred, the lizards will still be active at lower temperatures).

A low preferred temperature has been demonstrated in the tuatara, a unique lizard-like reptile which lives only on small islands on the east coast of North Island and in the Cook Strait between North and South Islands, New Zealand.

50

Interest in the tuatara has always centred on its unique anatomy, because it is the sole survivor of a group of reptiles which flourished millions of years before the age of the Dinosaurs. The first accounts of living tuataras were based mainly on short day-trips to the islands and these were misleading because the tuatara is nocturnal. It has a low preferred temperature and comes out to feed at temperatures as low as 7°C. Toleration of low temperature seems to be an adaptation to nocturnal activity but the advantage of such a regime is not at all obvious, except that the snails, beetles, crickets and earthworms which make up the tuatara's prey are also nocturnal.

The endothermic birds and mammals mark the final stage in independence from environmental conditions. Within limitations, their activity is not restricted by fluctuations of humidity or temperature so that this aspect of the night scene holds no particular advantage or disadvantage, except under the extreme conditions of a hot and dry climate. Nocturnal habits would, in consequence, seem to be ordered mainly by pressures of predation and the availability of food, so the birds and mammals can be quickly dismissed from a chapter devoted mainly to the physical conditions leading to a night life.

Birds are almost entirely diurnal, a fact which makes the few nocturnal representatives described in the next chapter particularly interesting. Mammals, on the other hand, are well represented at night. The roots of this fundamental difference lies in the early evolutionary history of the birds and mammals. They are descended from different branches of the reptile stock. The birds arose from reptiles which, it is thought, took to the trees to avoid larger, carnivorous reptiles and to feed on the insects which were swarming in the trees. They learned first to glide and then to fly. From the start, they used their eyes for the control of flight and searching for food and a dependence on vision has remained with the birds thereafter. The prototype mammals avoided their reptilian adversaries through another strategy. They took advantage of their warm-bloodedness to hide during the day and come out at night, when the cold-blooded reptiles slowed down or became inactive. In the darkness, the mammals developed the senses of touch, and especially smell, to guide them. Many mammals have retained nocturnal habits and a reliance on the sense of smell, which is used for communication in their social lives as well as for detecting food and danger. Man is an exception on both counts.

Of the main groups of mammals, the insectivores, bats, opossums, prosimians, armadillos and pangolins are primarily nocturnal. The squirrels, and the monkeys and apes, are almost the only strictly diurnal mammals, and they have their exceptions in the flying squirrels and the night monkey. Most other mammals are active through the 24 hours, either in bouts of activity and rest, or modifying their behaviour to suit local conditions. Shrews, small carnivores such as weasels, many rodents and hoofed animals feed in bouts of two or three hours because their nutritional requirements demand a frequent supply of raw material, but imposed on this pattern is a 24-hour cycle which results in more bouts occurring either by day or night. Shifting activity periods within the 24-hour cycle is flexible, as hoofed mammals adjust their activity to times when predators are less of a threat and predators adjust theirs to the time that gives them the advantage. Everywhere, mammals retreat into the night to avoid diurnal man.

Birds of the Night

Birds are an unmistakable group of animals with their wings and beaks, and suits of feathers. They also rank, with the well known and familiar exceptions of the owls and nightjars, as the most diurnal of animals and rely heavily on the sense of sight in all they do. Compared with amphibians and mammals, for instance, they are indeed predominantly active by day but a closer examination of the world of birds reveals night-fliers which are usually overlooked. Some of these are habitually nocturnal while others are active by night only at certain times. A survey of such nocturnal behaviour as there is in animals otherwise basically diurnal should shed some light on the reasons why they and other animals have opted to be active at night.

It is because we shut ourselves indoors at nightfall and later retire to bed that we fail to appreciate how often birds sing at night. The volume of song may be no more than a fraction of that heard by day, yet it is the case that robins, blackbirds and thrushes occasionally wake up and sing for a while, especially when they have been disturbed. Indeed, these snatches of song are sometimes mistaken for the outpourings of the nightingale, and if the name of this bird means nothing else it is a reminder that while we sleep there are more than merely owls performing under cover of night's dark mantle.

The very name 'nightingale' is Old English for night-singer. Poets have called it philomel, lover of song, for its powerful, deliberate phrases, with harsh and sweet piping notes intermingling. They are so enchanting to the ear on a warm, still, moonlit night that they have made the name of the bird synonymous with sweet singing. Thus, Jenny Lind, the opera singer, was nicknamed the Swedish Nightingale (and frogs have been sarcastically called Dutch nightingales!) Nightingales were taken to New Zealand years ago and they have been introduced into Florida for their song where, ironically, the mockingbirds have learned to mimic them. Yet many maintain that the song is overrated and claim that the nightingale owes its popularity to having no competition at night and that it is the atmosphere of the night that enhances the effect of its song. What is so often overlooked is that nightingales also sing by day when they are but one component of the chorus, competing with robins, blackbirds, thrushes, wrens and the warblers. Shakespeare summed it up.

Waders feed on the mudflats long after the sun has set. They retire to their roost only when the tide has flooded back.

'The Nightingale, if she could sing by day
When every goose is cackling, would be thought
No better a musician than the wren.'

53

A flock of cattle egrets roost in the safety of a tree, but they remain alert to disturbance and may take flight.

In fact, Shakespeare was not wholly correct as anyone will know who has had the rare and delicious experience of hearing a nightingale in full voice when, on a warm morning in May, there is a temporary pause in the full chorus of bird song and the superb voice of the nightingale is heard on its own, singing as it seldom sings after dark. In spite of all the traditional brouhaha about the nightingale, however, we are still no nearer an explanation why the bird should so persistently sing at night.

Leaving aside the owls, the nightjars and the nightingale for the moment, there are two other birds commonly heard in the European countryside at night. These are the snipe and woodcock. Both are waders, a group of birds more usually associated with the sea shore and sometimes named 'seashore birds'. Snipe, unlike the typical waders, frequent damp meadows, marshes and moorlands and their calls are highly evocative of long summer evenings, especially in the northern parts of its range, where the 'gloaming' of the Scots and the 'simmer-dim' of the Shetlanders stretches almost to midnight.

The snipe's 'chip-per, chip-per', delivered with the regularity of a metronome, carries well over fields and moors. The bird also has a second call, if the term may be permitted. This is a mechanical sound made by air rushing past two stiffened tail feathers, so that the snipe, as some people say, sings with its tail. In courtship, the male snipe dives sharply earthwards, with wings fluttering and the

54

The nightingale is famous for singing at night but some people find that the continuous medley of notes upsets their rest.

two special tail feathers standing out at right-angles to the tail. The slipstream sets the two feathers vibrating to produce hollow beating notes called drumming, sounding rather like someone blowing in short puffs across the mouth of an empty bottle. The effect is decidedly eerie when heard for the first time, largely because it is difficult to place the source of the sound, which has a ventriloquial quality, and to pick out the silhouette of the snipe against the darkening blue of the evening sky.

The woodcock is a woodland bird and it is most often seen flying in clearings and over woodland tracks. At dusk and dawn, daily from March to July, the male woodcock patrols his territory and advertises his presence with a display called roding. He flies in a fast circuit at tree-top height, or at about 10 metres over open ground, with a fairly slow wingbeat, calling with a thin, double 'tsi-wick'. At intervals he checks his forward movement, 'marking-time' for a couple of wingbeats, and croaks quietly.

One explanation that has been given for the nightingale's behaviour is that nightingales migrate at night and, therefore, arrive at their breeding grounds in the dark. The males arrive first, so, it is suggested, night-singing will attract the incoming females. This hypothesis lacks real foundation since there are other night migrants who are nonetheless wholly diurnal in their breeding habits and it does nothing to explain why nightingales continue to sing at night after they have

The snipe probes damp soil for food mainly at dawn and dusk.

Opposite top: The woodcock's plumage blends superbly with the woodland floor. As evening falls it will leave its nest and search for food.

Opposite bottom: The kiwi is another bird that probes the soil by night. It is aided by a good sense of smell and long 'bristly' feathers which act like a mammal's whiskers.

paired. Snipe and woodcock are different. They are more generally nocturnal, or, more correctly, crepuscular. They feed mainly after dusk, probing the soil for earthworms and insects with their long, sensitive bills. Their day is spent roosting quietly on the ground where their brown mottled plumage provides an excellent camouflage. Their nocturnalism seems linked wholly with feeding, their prey being most active at night, and this is emphasised by the way they are forced to feed more during the day when the ground is frozen.

A parallel from the opposite side of the world is seen in the behaviour of the flightless kiwi. It is a most unusual bird in many respects. The tail is lacking, the tiny wings are hidden under the loose coat of long, hairlike feathers and the bird's ungainly appearance is completed by the long slender bill. Although familiar as an emblem of the New Zealanders, the kiwi is not often seen in the wild. It leads a retiring life in the damp forests of New Zealand, where it is active at night. Unusual for a nocturnal bird, it has small eyes, but its sense of smell is keen. Kiwis feed largely on earthworms which they find by probing into the soil with the long bill. The tip is sensitive to touch but the nostrils are placed near the tip and experiments have demonstrated that kiwis can find their food by smell.

Although less obvious than the nightingale, snipe and woodcock which noisily notify their territorial claims, many water and shore birds actively feed by night as well as by day. Most geese, as well as swans, are mainly diurnal and retire to roost at night, but the pink-footed and white-fronted geese continue to graze at night when there is a moon. Because the brent goose forages more in the shallow water of estuaries and mudflats its feeding has to be geared to the cycle of the tides, and it also seems to take advantage of moonlight to feed when the tide is ebbing at night. Among ducks, the wigeon's activity is also regulated by tide and weather but it retires to roost in large flocks on open water during the day

56

when there is disturbance. Rafting, as this is called, on open water by day is commonly employed by other ducks as a defence. Large numbers of teal, for instance, gather for the winter in the Camargue and there they feed in the shallow water of the marshes at night. By day, they form flocks on pools as a defence against marsh harriers. The common mallard, by contrast, is flexible in its behaviour yet even it feeds mainly by night in towns where it is subject to disturbance during the day.

The waders which feed on small invertebrates extracted from the mud and sand of shore and estuaries have a pattern of feeding and roosting which is governed by the tide. They spread out over exposed beaches and flats while the tide is out, and move upshore as the sea returns. When the tide is high, the waders roost, often in dense flocks. This pattern of feeding and roosting continues throughout the 24-hours but the level of activity at night is increased by the amount of moonlight available.

These examples, at least, show nocturnal behaviour among birds as a response to the availability of food, as in the uncovering of food at low tide for waders, or to the avoidance of disturbance and harassment. Another example of the latter is found in flamingos. They are essentially diurnal in their habits but they fly from lake to lake at night. Although strong fliers, they are not agile and form an easy target for birds of prey. It is clear, therefore, that even diurnal birds can invade the night, either to feed or to fly.

The waders, geese and ducks find their food largely by 'feel', so light is not essential, and we must presume that roding by woodcock and the drumming of snipe, as well as the movements of flamingos take place when it is not too dark for the birds to see possible obstructions. It would be a different matter if the birds had to find and capture moving prey or navigate through trees, and this is the situation facing owls and nightjars.

The owls, with the large, round head, forward-facing eyes and sober bearing, are an easily recognised group of birds found all over the world, in every sort of habitat from forests to deserts. Of the 130 or so species of owl, about two-thirds hunt mainly in twilight and darkness. In contrast to the few hawks, eagles and falcons that have invaded the twilight, a fair number of owls are active in daytime. Of these, the snowy owl, Tengmalm's or the boreal owl and the great grey owl, among others, are compelled to hunt in daylight during the Arctic summer when the sun never sets. The burrowing owl of North American deserts, the pigmy owls, the hawk owl of northern North America and Eurasia and the Australian boobook and barking owls go further. These are habitual day hunters and there are yet others that may hunt by both day and night. Barn owls and short-eared owls are not uncommonly seen quartering the ground in bright sunlight, and even the tawny owl, usually considered to be the most nocturnal of all owls, can be heard hooting during the day or occasionally seen flitting silently through the trees.

Nevertheless, the owls as a group are the nocturnal counterpart of the day-flying raptors, to the extent of there being complementary species hunting over the same ground. In North America, for instance, the great horned owl hunts hares by night and the red-tailed hawk hunts them by day; the short-eared

Although sometimes seen abroad by day, owls like this long-eared owl are the nocturnal counterparts of the hawks and falcons, with whom they share the hooked, tearing bill and sharp, grasping talons.

owl replaces the hen harrier or marsh hawk; the barred owl replaces the red-shouldered hawk and the screech owl replaces the sparrowhawk or kestrel. A neat demonstration of time-sharing takes place in the Galapagos Islands where there is one owl and one hawk. On islands where the short-eared owl lives alone it hunts by both day and night, but where the Galapagos hawk is also present, the owl is confined to the night. Should it appear by day it is immediately attacked by the hawk.

The success of the owls rests on the development of two senses—vision and hearing—which enable them to hunt at night. There had been speculation over the years as to how an owl is able to pinpoint its prey even in pitch darkness, and only in recent years has this almost uncanny ability been satisfactorily explained. At one time it was claimed that the eyes of owls are sensitive to infra-red radiation, so they could see in what we call pitch darkness. Then, some experiments seemed to show that owls could catch mice in such dim light that their eyes would have to be up to 100 times as sensitive as ours. These experiments consisted of releasing barn owls in a room to find dead mice on the floor under conditions of diminishing light intensity. In fact, the aim of the experiments was to investigate the camouflage value of mouse fur, not the visual sensitivity of owls, but the results were used to claim supersensitivity for owl eyes.

There is no doubt that owls have good eyesight. The eyeballs are huge, as big as human eyes in some species, and they are squashed into the skull. They are tubular, as are the eyes of nocturnal bushbabies, so that a large lens and retina can be fitted into a limited space. The forward-looking position of the eyes is necessary to accommodate the large eyeballs and, as there is room to revolve no more than a fraction in their sockets, the owl has to turn its head to stare at anything that takes its interest. Owls are famous for their ability to rotate their heads until they are looking over their backs and to turn their heads almost upside down.

The large cornea and lens, coupled with an iris which dilates widely at night, allow the eye to gather plenty of light. In terms familiar to photographers, a tawny owl's eye has an f number of 1.3, compared with f = 0.9 for a cat and f = 2.1 for human beings. Although the retina has more rods than cones and is provided with a tapetum, its overall sensitivity is little different from that of the human retina. The difference in night vision between owls and men lies in the owls' better light-gathering capacity.

Following careful investigation Graham Martin concluded that owls' eyes are no more than $2\frac{1}{2}$ times more sensitive than human eyes. To put this in perspective, he points out that a day on a sunny beach can reduce our sensitivity three times, so that a difference of $2\frac{1}{2}$ times is hardly significant. The important comparison is between an owl and a pigeon. A pigeon eye has an f number of 4.0, which represents a considerably reduced light-gathering ability and an overall sensitivity in the order of 100 times poorer than the owl. Pigeons probably have poorer night vision than many other birds. They always go to roost before dark and are never active at night but the comparison between pigeons and owls does, however, indicate the advantage owls have, especially in twilight or moonlight. Nevertheless, the success achieved by owls at hunting under trees on dark nights has another explanation.

When the light is too dim for them to pick out their prey visually, owls rely on their ears which are capable of pinpointing faint rustling movements. This capability was demonstrated by a test which more successfully involved barn owls, mice and a darkened room than the experiment which was thought to have demonstrated superlative visual ability. The barn owls pounced accurately on a mouse scuffling over the leaf-strewn floor of the completely lightproof room. They would also strike at a ball of paper pulled by a string through the leaves, but not at a mouse running noiselessly over a bare floor under the same conditions. Finally, it was found that the owls could locate the sounds of leaf rustles coming from a loudspeaker. If the rustles stopped after the owl had left its perch, it would still strike home, so the owl, in contrast to a guided missile which follows a target, directs itself at its prey and then flies as straight as an arrow.

Owls locate the direction of sounds in the same way as we do. A sound coming from one side will take a fraction longer to reach the farther ear and will have a slight difference in frequency pattern. It will also sound a fraction louder in the nearer ear. The nervous 'computer' in the brain compares signals coming from each ear to plot the position of the sound source. The system of comparing time differences only works for sounds which have a wavelength shorter than the

An eagle owl's plumage disguises the contours of its body. Under the rounded shape of the head, there is a pair of ultrasensitive ears. The tufts of ears are ornamental and have nothing to do with hearing.

60

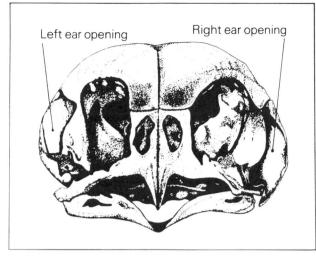

Left ear opening   Right ear opening

Above: A drawing of an owl's skull shows how the ears are asymmetrically placed. This helps the owl to locate the source of the sound. (*After* Burton, *Owls of the World*, Peter Lowe).

Right: A fledgling tawny owl faces a stern task. It has to learn the skills necessary to make a living by hunting at night.

Opposite: Two rapid series of photographs showed how a barn owl catches its prey. In the dark, it flew with its head pointed at the target and feet dangling. At the last moment, it flung its talons forward, where its face had been. In the light, it glided down and swung its talons down as it neared the target. (*After* Payne, The Living Bird 1, 1962).

distance between the ear openings, which means that the animal, such as an owl, will be able to locate sounds of higher rather than lower pitch whereas man, with his larger head, is able to locate sound of lower pitch than can an owl.

Sensitivity in the owl ear is improved by refinements in the internal mechanisms and also by a flap of skin, which like the mammalian pinna, helps to channel sound into the ear. In some owls, for example the barn owl, tawny owl and eagle owl, one ear opening and its flap is placed higher on the head than the other, thus helping to locate sounds with greater accuracy. As a mouse runs past, the sound of its movements changes more rapidly in one ear than the other so that the differences between the two are magnified.

It would be a reasonable assumption that owls hunting by day or in twilight use their eyes to guide their attack, even if sounds alert them to the presence of their prey. Further, hearing may be the only guidance when it is too dark for the bird to see properly and photographic records have indeed been used to demonstrate that barn owls employ different tactics according to light intensity.

The plumage of an owl gives it an extra advantage over its prey. The feathers have fluffy edges which reduce the noise of flight. This is a great horned owl.

If there is sufficient light, the owl glides down on outstretched wings and, at the moment of striking, throws up its wings and swings its talons forward to grab the prey. When it is very dark, the owl flaps down with its head pointing undeviating at the source of the sound and legs dangling. Then, in the split second before impact, the legs are swung forward and the talons placed where the face had been a moment earlier, in an elegant display of muscular and nervous coordination.

Owls hunt either by patrolling on the wing, quartering low over the ground, or by waiting on a convenient perch, and in both cases drop on the prey with a sudden pounce. The small mammals which are the main prey of many owls have equally sharp senses to warn them of impending attack. At night, they rely on hearing an owl's approach and the owls have countered this by developing muffling to deaden the sound of their flight. Even a small bird makes a surprisingly loud whirring with its wingbeats but the owls are renowned for their silent flight. They are relatively light-weight for the wing area, so their flight is buoyant. The easy flapping of their wings will generate less noise than the laboured beating of a heavier bird, and, anyway, the final attack is a quieter glide. The noise of passage is further deadened by a fine velvety fringe around the edges of the flight feathers. They are the owls' equivalent of the sacking swathed around oars to muffle a boat's progress through water. This muffling must make flying more strenuous and it is of interest to note that the fringes are missing from the wings of purely diurnal owls, who do not need them.

This nightjar shows the large eyes it uses for hunting flying insects and the white spots on its wings which are used for identification in dim light.

On the whole, owls rarely hunt flying animals. Surprisingly few catch bats, although the Australian hawk owls are an exception. Some owls, like the elf, little and pygmy owls, feed largely on insects which they catch on the ground but a few, including the New Guinea hawk owl and the scops owl, hunt on the wing.

Flying insects are the speciality of the nightjars. They and their relatives comprise a group of about 95 species of birds. They share the common features of large eyes, a wide, gaping mouth and a plumage of mottled browns. These characters reflect the habits of the nightjars. Most are nocturnal, feeding on flying insects, and they spend the day crouching on the ground or perching on a branch, with the plumage acting as a camouflage to render them extremely unobtrusive. As a consequence nightjars are usually seen when caught in motor car headlights and they are frequently killed because they have been resting on the roadway. This is particularly true in the tropics, possibly because of the flying insects that gather over the roads.

The 70 or so true nightjars are found over much of the world, except the far north and New Zealand. They live in open country of heaths, moors and dunes, and on the savannahs of Africa. The name nightjar reflects their harsh churring calls, although in North America they may be called nighthawks or goatsuckers, the last name recalling the folk belief that they drink milk from nanny goats.

The monotonous calls betray the nightjar's presence and give the names of two American species. The poor-will of western North America, which is famous as the first bird proved to hibernate, has a monotonous call, apparently of two notes rendered 'poor-will' but, at close quarters, a third, quiet, note can be heard and the bird would be better named 'poor-will-low'. Similarly the whip-poor-will should be the 'whip-pup-poor-will', although the second syllable can be heard only by recording the call and playing it back at slow speed.

Nightjars are often described as crepuscular rather than nocturnal, but I suspect this reflects the habits of ornithologists rather than of nightjars as they can be heard throughout the night. There are, no doubt, peaks of activity in the twilight hours and we would expect nightjars to be most active when there is sufficient light to pick out flying insects easily. Insects are also more active in the early part of the night. Moonlit nights must induce nightjars to continue feeding through the night, while dark, overcast conditions dampen their activity. As the calls show, nightjars' night activities are not limited to feeding; their social life also takes place in twilight or darkness.

Nightjars, as a group, make great use of visual displays for courtship and threat. The male European nightjar has two or three white spots on its wings and one on each corner of the tail, which the females and juveniles lack. The whip-poor-will, common nighthawk and other species have their own distinctive patterns of white spots which must make the males' displays more visible as well as instantly identifiable. The pennant-winged and standard-winged nightjars of Africa have trailing primary wing feathers, each ending in a small flag. A male pennant-winged nightjar flies around the female with these feathers held erect like the lances tipped with pennonsels of mediaeval knights. These are readily visible to the female since the courtship is carried out before full dark of night.

It is sometimes said that nightjars catch insects pretty much at random, by sailing through the air with mouths wide open like a fishing trawl. Such a chance method would be very wasteful in time and energy and would probably not work. Water or air flows through a net and solid objects are trapped on its meshes, but a bird flying with its mouth open will be no better at catching insects than a bucket is for catching fish. The quarry will be deflected away.

When feeding, nightjars exhibit the same twisting and jinking flight as swallows and bats, and give the distinct impression of chasing and snapping individual insects. The European nightjar most commonly takes large insects, such as moths, beetles and craneflies, which are easy to pick out in dim light. The catching area of the wide gape is extended by a fringe of bristles around the mouth, so the nightjar need not rely on pinpoint accuracy in snapping up an insect. However, the oilbird, a near relative which feeds on fruit, still possesses bristles so they may have a tactile function like a cat's whiskers.

An alternative to hawking through the air in search of insects is to use a perch as a vantage point, the bird letting the insects do the travelling and itself doing no more than make a short sally to snap them up. Such is the strategy of the fiery-necked nightjar of Southern Africa and the potoos of Middle and South America. In Australia and the Orient, the frogmouths use a perch to launch attacks on small animals passing underneath, catching large insects, scorpions, frogs, even small mammals and birds. The owlet frogmouths employ both hawking and sallying, but take most of their food from the ground. Such feeding habits make these species the counterparts of the swallows, swifts, bee-eaters, flycatchers and shrikes, and competition is avoided by their nocturnal habits.

66

Neither do their diets overlap with owls, and the owlet frogmouths, which look like owls with large mouths, are confined to Australia and Papua, where there are few competing small insect-hunting owls. Competition with bats must be another matter for nocturnal birds, and I have seen a nightjar narrowly miss colliding with a serotine bat, which was presumably chasing the same insect.

Northern South America, including the island of Trinidad, is the home of the oilbird, a relative of the nightjars which is unique in several respects. It is so named because the young used to be taken from the nest for the sake of their fat which renders down to a fine cooking oil. Unlike nightjars and related birds, which nest on the bare ground or in trees, the oilbirds nest on ledges in caves. They find their way to their nests in pitch darkness by emitting a series of echolocating clicks, which are audible to the human ear, but the echolocation capability is crude compared with the sophisticated system of bats.

Oilbirds emerge from their caves in the evening, to the accompaniment of eerie shrieks and screams which gave rise to the belief that oilbirds house the souls of criminals. Their echolocation is not used in the open, the birds navigating by sight as they search for palms and other fruit-bearing trees. Oilbirds are unique in being the only nocturnal bird to feed on fruit, some of which are strong-smelling and it is thought that oilbirds use a sense of smell to help locate them. If this is so, the combined use of the senses of vision, smell and echolocation is another unique feature of these strange birds.

The albatrosses, petrels and shearwaters are deep-sea wanderers. Known collectively as tubenoses because of their horny tubular nostrils, they quarter the seas far from shore, using the wind to fly economically in their search for food. The giant albatrosses, the largest of the group with wingspans of up to 3 metres or more, are famed for their ability to glide for minute after minute without flapping their wings. At the other end of the scale the storm petrels look more like large house martins. Between the two range the many kinds of fulmar, shearwater, petrel, prion, diving petrel and gadfly petrel.

The diet of these birds is composed of small marine animals, especially fishes, squids and crustaceans. Because of the difficulties of studying birds feeding at sea, our knowledge of their habits has been studied mainly by catching the birds and examining the contents of their crops or by watching them as they fly by the ship. As will be described later, many of the animals on which the tubenoses feed migrate to the surface at night and the inference is that the birds feed mainly at night. With the exception of the diving petrels, which use their wings to 'fly' through the water, few of the tubenoses can dive well and they are restricted to feeding off the surface or plunging no further than the uppermost layers.

Finding prey swimming near the surface by night and then attacking it accurately poses problems. Marine creatures will not show up from above in the same way that flying insects may be picked out against the paler backdrop of the sky. The exception is those organisms which bear light-producing organs. We can only assume that nocturnal seabirds have eyesight and feeding habits which can overcome the problem of darkness and enable them to pick out their prey in the dark, but it has been known for many years that petrels and their relatives have an unusually well-developed sense of smell. This was established by dissection which showed that the sensory membranes in the nose and the associated nervous tissues are better developed than in the majority of birds.

Moreover, petrels are attracted to oily or fatty substances floating on the sea. Dark-rumped petrels of the Galapagos and Hawaii used to be caught by luring them within gunshot with turtle fat. More careful experiments have consisted of floating fish or vegetable oils within a flotation ring or impregnating a sponge on the end of a short pole mounted on a buoy. Albatrosses, shearwaters, fulmars and storm petrels were attracted to these baits but terns, gulls, gannets and puffins ignored them. Typically, the tubenoses approached the baits from downwind, flying in low zigzags over the water, like a bloodhound quartering the ground. Sometimes they would make several passes, as if firmly convinced that there was food there but fooled as to its exact location.

The oceanic tubenoses come ashore only to breed and they nest in colonies, sometimes of thousands of pairs, on cliffs and islands. A few species, like the dark-rumped petrel and the cahow of Bermuda, frequent tropical regions but the group is most abundant in cooler seas and their main home is the temperate and sub-Antarctic regions of the southern hemisphere. The remote nature of the colonies offers the birds protection from predators. Like many birds which are adapted for an aquatic life, the tubenoses are ungainly on land. The legs are placed well back on the body for efficient swimming but this leaves the bird unbalanced on land and an easy prey for terrestrial hunters. Oceanic islands have very few, if any, native land predators and the seabird colonies were safe until man unintentionally or deliberately brought in mammals from other parts of the world. Feral dogs and cats, pigs ranging freely, rats from ships, the mongooses

which were introduced to control snakes in the West Indies, have all posed serious problems in many places. The birds have no defence against these. In a natural state they are threatened only by other birds and their sole defence from these is to visit their nests by night. A visit to a colony of tubenoses by day reveals little or no sign of occupation, but it comes to life at night. On bright nights, the birds may be seen flying in and out or wheeling over the colony, but on the darkest night all the visitor is left with is the noise of the birds calling from their nests or on the wing.

The Manx shearwater nests on many islands of the Pacific, North Atlantic and Mediterranean, although it is no longer found on the Isle of Man, the island after which it is named. It appears only on the darkest nights when it fills the air with a varied chorus of gurgles, screams and wails, which have been described as sounding like a cross between a rooster with a sore throat and a baby being strangled. The need to visit their nests and display in flight over the colony only on the darkest nights is amply demonstrated by the onslaughts of great black-backed gulls on any shearwaters which happen to come ashore at dawn or dusk or in moonlight.

My first experience of tubenosed birds came during a stay on an island on the fringes of the Antarctic. Five species of petrel nest in the slopes and cliffs around the scientific station there, and after I had arranged for our cat to be repatriated, the only enemy of the birds was the skua. This is a predatory member of the gull family which uniquely breeds in both north and south hemispheres. Once a week I walked around the petrel colonies and visited skua nests, collecting the remains of birds which had been caught on the ground and pecked to death.

Three of the petrels—Wilson's and black-bellied storm petrels and the dove prion—were very rarely seen around the island by day. And if they were, they were usually being chased by skuas. The rather larger snow petrel and cape pigeon, which resemble the fulmar of the North Atlantic, were, on the other hand, often in evidence, flying to and from their nests or circling the cliffs. The snow petrel nests in large crevices or deep, overhung ledges on cliffs but the cape pigeon prefers more exposed ledges. I believe that this nesting habit confers a certain immunity to attack and so allows a diurnal habit. The smaller petrels nest in burrows in soft peaty soil or under boulders, which would at first appear more secure. However, to enter its burrow a petrel must land a metre or so downwind and shuffle up to the entrance, and in this position it is highly vulnerable because it can take off only with difficulty. The cape pigeon and snow petrel are exposed on their ledges but they can fly straight on and off, avoiding the vulnerable shuffle, and the petrels' well known habit of spitting an evil-smelling, clogging oil at intruders protects them while they are on the nest. Neither system is a perfect defence as the attentiveness of the skuas and the litter of pecked corpses attests.

A second trip to the Antarctic regions took me to Bird Island, a small satellite of the large island of South Georgia. Bird Island is a seabird paradise and is the breeding ground of hundreds of thousands of birds ranging from the gaint wandering albatross to the comparatively minute storm petrels. The albatrosses, four species in all, and the two giant petrels lay their eggs on mounds of mud and vegetation. They arrive and depart at will and they are far too large to be bothered by skuas. Of the several burrow-nesting petrels, all except one are nocturnal in

70

Dove prions are members of the petrel group which live in southern oceans and fly unerringly to their nests when it is dark.

their activities on the island. The odd one is the white-chinned petrel which also seems to be too large for the skuas to deal with. We never saw one attacked and we never found their eaten remains. The constraint that the skuas put on entering the nest by day seems to be clinched by observations of more diurnal behaviour by small petrels living at latitudes farther from the polar regions where there are no skuas.

Where night-flying is necessary, the petrels have to find their nest burrows in the dark. After a heavy fall of snow, petrels in the Antarctic burrow through the freshly fallen snow with unerring accuracy to reach the entrance. The position of landmarks would allow pinpoint navigation but petrels are, as we have seen, more in evidence on the darkest nights. There may be however, sufficient light for the loom of cliffs and hills to guide the birds, but, as with finding food in the ocean, the sense of smell has been invoked to explain how petrels find their burrows in the absence of visual clues. Experiments with Leach's petrel, one of the storm petrels, suggest that it will home to the smell of its nest material and petrels with their sense of smell artificially impaired failed to return to their nests.

The comings and goings of seabirds around their colonies at night are witnessed only by hardy ornithologists willing to visit remote islands and coasts and give up their sleep to visit the colonies, but nocturnal movements of birds are much more general. That birds migrate at night has been known for years but the extent and regularity of their movements have been appreciated only in the last 30 years. Lighthouse keepers have always been aware of the danger to birds posed by the powerful beams which dazzle and attract thousands to their death. People out on autumn nights have long been aware of the call notes of birds passing overhead. For example, the thin, lisping *tseep* notes floating down from the darkened sky, like the whisperings of disembodied spirits, mark the arrival of redwings in Britain from their Scandinavian home. However, the significance of these manifestations was not appreciated until World War II when the night movements of birds became involved in national defence against air-raids.

It was a saying among the men of wartime Bomber Command that 'only birds and fools fly—and birds don't fly at night'. Evidence that the second part of this statement is incorrect was even then to hand. The radar sets which maintained the vital surveillance of the airspace were bedevilled by unexplained 'shoals' of small echoes nicknamed angels, which did not appear to have any physical manifestation. It eventually transpired that these angels were the radar echoes of migrating birds. This came as a surprise because nobody had thought that birds the size of finches and warblers could send back echoes. Once this had been appreciated radar became a very useful tool for studying the movements of birds by day and night. As well as showing the direction, speed and height of the birds, it is possible with radar to estimate their numbers, and the studies assumed a greater importance with the expansion of civil aviation and the increased danger of planes running into birds and crashing.

A simpler and more elegant technique of studying nocturnal migrants which is available to anyone, is to 'moon-watch'. A telescope, such as the × 20 spotting scope used by the keenest birdwatchers, is trained on the moon's disc and passing migrants can be seen as silhouettes. There are obvious limitations imposed by bad weather and a new moon but, with the aid of mathematical computations, moon-watching has yielded a great deal of information on the direction and volume of the migratory flow.

The study of bird migration is a wide, complex subject which embraces the many problems of navigation, the provision of energy for long, non-stop flights, the timing of departure and the puzzle of how a young bird knows when it has reached its destination for the first time. Even the reason for migration is still a matter for learned discussion and despite 30 years of research it is still possible to refer to the mysteries of migration. In the context of nature's night life, interest is centred on navigation in the dark and the value of travelling at night.

It is becoming clear that birds have several means of navigation at their disposal, as has the modern navigator. They may use the position of the sun by day, the patterns of stars by night, the Earth's magnetic field, a knowledge of topography and, it has been suggested recently, the smell of their environment. It is likely that birds use the sun to plot an exact position, measuring its altitude and azimuth like the human navigator with his sextant. Night migrants then fix their bearings during the day and set off on a compass course. They could use the moon, but many birds travel when the moon is below the horizon. The most likely means of maintaining a correct heading at night involves using the patterns of stars overhead although it has been found that migrants continue to hold their course when the stars are obscured.

If the weather is clear, the coastlines, the lie of the land, or the set of waves for oceanic migrants, give a reference to prevent the birds being blown off course by side winds. A complete clampdown of the weather is not a disaster because a

The famous flamingos of Lake Nakuru in Kenya fly to their roost on nearby Lake Hannington in the evening to avoid attacks by predatory birds.

magnetic sense, which has been demonstrated in European robins, may keep the bird heading in the right direction. It does not, however, allow the bird to compensate for being blown downwind and it is not surprising that migrants are unwilling to set off in bad weather, although they will keep going if the weather deteriorates while they are airborne. When it becomes so bad that nothing can be seen or the wind is too strong for the birds to compensate, they land until conditions improve. The position can be checked and a new heading selected to bring them back on course. How birds recognise star patterns or determine the orientation of the Earth's magnetism remain mysteries. It is sufficient for our purposes to note that experiments have demonstrated that birds can use them to guide their movements and it is interesting that they do not rely on a single steering system but can switch from one to another as circumstances dictate.

The bomber pilots' adage that relegates night-flying to fools implies that navigation and recognising the destination on arrival is difficult at night. This is as applicable to birds as to aircrews, and there has to be good reason to fly at this time. British bombers made sorties over Germany by night to avoid predation, so to speak, and migrant birds must benefit from the absence of diurnal birds of prey, such as Eleanora's falcon of the Mediterranean and the sooty falcon of eastern Africa which nest late in the summer so that they can feed their growing young on the flocks of small birds heading down to southern Africa by day. This is not the whole explanation and, as is so often the case, there is no clear picture of why birds choose to migrate by day or night.

Some birds must of necessity migrate by both day and night. The waders and wildfowl which travel long distances over the sea may have to remain airborne for over 12 hours and they will then experience both sets of conditions. Regular diurnal migrants include the birds of prey, pelicans, storks, swifts, swallows, many finches, starlings, members of the crow family and turtle doves. The night is preferred by thrushes, cuckoos, flycatchers, warblers and buntings.

Among the smaller birds there is a rough division of insect-eaters travelling by night and seed-eaters by day. Migration at night allows the insect-eaters to spend the day foraging for food. This is not so necessary for seed-eaters because their diet enables them easily to accumulate the necessary reserves for the long flight. The notable exceptions to this scheme are the day-flying swifts and swallows, but these birds hawk for insects in flight and they can continue feeding as they travel. The typical pattern for a nocturnal migrant is to replace the time normally spent roosting with travelling and to leave the day clear for feeding.

This explanation does not deny the immunity to predation conferred by the night, and physical conditions may also be more favourable at night, especially in the tropics. Many of Europe's smaller migrants have to make the long crossing of the Sahara on their journeys to and from their wintering grounds. The night gives a more equable environment to small birds which might find the effect of a blistering sun an added strain on bodies already burdened with the heat produced by hours of flying. Moreover, thermals are bubbling up from the desert surface during the day making the atmosphere seethe like the fat in a frying pan and small birds would be buffeted and would waste energy struggling to keep on course.

Bats

The emergence of streams of bats from huge underground roosts is one of the wonders of the animal world. It can be witnessed at scattered points around the warmer parts of the world, where large caverns provide safe refuges for bats. There are the Batu Caves of Malaysia, the Niah Cave of Borneo, the Tamana Cave of Trinidad, and the Carlsbad Caverns of New Mexico. The Carlsbad Caverns were the home of an estimated 9 million bats, until their numbers declined in recent years. They issue from a single entrance at dusk in a continuous black stream which can be seen several miles away and it used to take a full 20 minutes for all the bats to emerge in a column so dense that there was continuous rattle and flutter as the bats' wings struck each other.

The bats are renowned for several reasons. They are the only true flying mammals; they hibernate in caves, buildings or tree crevices; they are nocturnal and they have a system of ultrasonic echolocation for detecting and tracking their prey. However, the numbers and varieties of bats are not often appreciated; the 950 species make the order Chiroptera—the bats—the largest group of mammals after the rodents. While bats are most familiar as insect-eaters, many tropical species eat fruit, pollen and nectar, a few hunt small animals such as other bats and fishes, and the famous vampires live on blood. Bats are so little known that few have received common names. They are all too often associated with witchcraft and other nefarious nocturnal activities or distrusted because of their association with the appalling disease of rabies. Yet by virtue of their unique qualities and variety of habits, the bats are one of the most interesting, although most neglected, kinds of mammal. Here we are interested in those aspects of bat biology which have enabled them to exploit the night and to see what are the advantages of night life for bats, as a contrast to the basically day-flying birds.

There are two main groups of bats: the Microchiroptera and the Megachiroptera. The latter comprise the fruit bats and are entirely tropical in their distribution; the Microchiroptera are mostly insect-eaters and mostly tropical but they are also common in temperate countries.

The lack of common names makes it difficult to appreciate the different kinds of bats and this is made no easier by the overall similarity of physical appearance, although they are no different in this respect from the rodents. The latter are, however, more familiar and the variations in faces and tails which are used to distinguish families of bats mean little when their owners are so unfamiliar.

75

Birds and bats together. Swiftlets mix with bats near the entrance to their roosting cave in Sarawak.

It is rare to see a bat flying in broad daylight, but it sometimes happens when, for example, cold weather has prevented the bat from obtaining enough food at night. Normally, the time of emergence is closely synchronised with the time of sunset. The bats emerge from communal roosts in a fairly steady and readily visible flow about an hour after sunset. Bats roosting alone also time their departure but they are less easy to watch. At one time I had a pipistrelle roosting under my guttering and I could stand at the window and watch it fly off into the evening without too much waiting. On fine evenings it is still light enough to see the bats clearly although it takes an expert to do more than hazard a guess at their identification. The bats time their departure with accuracy by making brief sorties from the depths of the cavern or roof space or by loitering at the entrance where they can check on light intensity. They also have a natural circadian rhythm which is altered to fit the changing sunset by visual inspections.

After they have left the roost, it is difficult to follow the bats unless they can be tracked to a favoured feeding ground, for instance over a rubbish tip which has attracted hordes of flies, or a lake or woodland ride, where individuals patrol to and fro. The feeding habits of bats have, therefore, been something of a mystery, except that their diet can be analysed by picking the hard, indigestible fragments of insect bodies from droppings. The subject is now being studied by equipping bats with miniature radio transmitters and tiny flashlights so that they can be followed on their nightly travels and kept under observation with the assistance of night-viewing devices. As might be expected, particular species have their

76

Noctules usually fly high and fast in search of flying insects but they also chase prey on the ground.

individual foraging habits; some hunt only in open woodland, while others can change from woods to pastures as the seasons change.

Overall, the greatest activity by the insect-eating bats comes during the first half of the night, which is the time when flying insects are most abundant. Some species continue to fly through the night but others return to roost through the midnight hours and emerge again for a second bout of feeding, before returning to the roost before sunrise.

It is not normally possible to see how a bat catches an insect, unless the bat is out by day, but it has been found that insects are mostly caught in the mouth, although a wing may be used to 'field' the insect and direct it towards the mouth. Insects may also be caught in the tail membrane, which is brought forward to make a pouch, and while still flying, the bat bends its head into the pouch to deal with the insect. Each insect is chased and snapped up singly; bats do not fly around with their mouths open on the offchance that something will be caught; their catch rate is too high for such an inefficient method.

Pipistrelles can catch a quarter of their own body weight of insects in 30 minutes. A captive little brown bat has caught tiny fruit flies at the rate of one every three seconds. However, not all insects are caught in the air. Long-eared bats, among others, hover among the foliage of trees and pluck caterpillars and spiders as well as resting winged insects. Others land on the ground to pursue grasshoppers and beetles. By walking on their 'knuckles' and holding the wing

77

Long-eared bat hovering. It often feeds by plucking insects from foliage.

membranes clear of the ground, these bats can scuttle along with surprising agility. The pallid bat of America pounces on its prey, which includes scorpions, while flying low over the ground. The large false vampires of America, the largest bat in the continent, capture geckos and bats.

The way in which bats locate and catch their prey and find their way in the dark has always been the feature of bat biology to excite the most interest. The naturalist W.H. Hudson tried to frighten two pipistrelle bats which flew to and fro over his head as he walked up a country lane. He spun his thin cane around so that it described a cone above his head and, to his surprise, the bats continued to fly over him without being thumped. It has been known for two hundred years that bats can fly without crashing into things when blinded but not when their ears were blocked, but how they did so remained a mystery. There were suggestions that the sense of hearing was involved, even that they were employing echolocation, but the system of ultrasonic echolocation could not be demonstrated until the development of suitable apparatus for detecting sound too high-pitched for human hearing. This came in 1938 and since that time research has shown bat echolocation to be extremely sophisticated.

The essence of echolocation is that the bat emits a continuous series of clicks and listens for the faint echoes which tell it the position of insect prey or obstacles in its path. The clicks are ultrasonic, which means that the pitch, or frequency, of the sounds is too high for the human ear. They range from about 15 to 150 kHz (1 kHz, or kilohertz, equals 1000 cycles/second) and, as the upper range

78

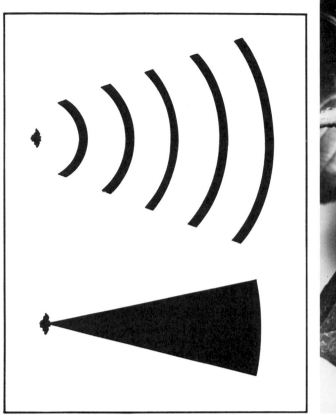

Two basic types of echolocation. The leaf-nosed bats emit a long, constant-frequency pulse. Other bats emit a stream of short pulses, each of which sweeps through a range of sound frequencies.

Right: A whiskered bat about to take off. It is already testing its surroundings with its sonar.

of human hearing is up to 20 kHz, the clicks can just be heard as a measured ticking by some people, especially young people.

The Microchiroptera employ two basic systems of echolocation. The majority 'shout' through their mouths, each click being a pulse of sound which sweeps down a wide range of frequencies in a glissando lasting no more than a few thousandths of a second and dropping through nearly an octave. The theory of radar shows that these short sweeps of changing frequency are ideal for measuring distances. This is called frequency modulation and as the bat patrols, it emits these clicks steadily at a rate of four or five per second. When it detects an insect, the clicking rate speeds up as the bat 'concentrates' on its prey and ends with a buzz of up to 200 clicks per second as it closes in.

The sensitivity of bat echolocation is superlative. Although their 'shouts' are so loud that we would be deafened if we could hear them, the echoes rebounding from an insect are weak in the extreme, and yet bats can pick them out from a clutter of background noise and a jumble of echoes from objects of no interest, with sufficient acuity to make use of them. They can avoid wires as fine as human hair and detect tiny fruit flies when they are 50 centimetres or more distant. The higher the frequency that the bat emits, the shorter will be its wavelength, and as objects reflect sounds with a wavelength less than their own width, high frequencies are best for locating tiny insects. The only drawback to using high frequency sounds is that they attenuate most rapidly, losing their energy to the air molecules. This problem is solved by 'shouting' as loudly as

79

The yellow-winged bat of Africa hunts like a flycatcher. It waits on its perch, then flies out to snap up an insect and returns.

possible. The energy of pulse can be increased by lengthening it, but a long pulse reduces the ability to distinguish between two objects. The sweep allows the bat to lengthen its pulses, and raise their energy and consequently the range of its echolocation, because echoes from two nearby objects will be reflected at slightly different frequencies and are so distinguishable.

The leaf-nosed bats, which include the horseshoe bats, false vampires, vampires and spear-nosed bats, employ a rather different echolocation system from the frequency modulation bats. The pulses are ten times as long and they are emitted as an almost pure tone of constant frequency, with a short sweep to finish. These bats are known as constant frequency bats. The sounds are produced by 'snorting' through the nose, instead of 'shouting', the bats possessing intricate flaps of naked skin on the snout, which gives them their common names. These 'nose leaves' act as a megaphone beaming a concentrated cone of sound, as opposed to the pulses of a frequency modulation bat which spread out like ripples. By twisting the nose leaves, the bat can scan ahead with its sound beam. The use of long pulses appears to run contrary to the need for brief pulses stated earlier, but these bats are using a different method of distinguishing echoes from different objects. Instead of measuring range and direction by taking a series of bearings in a straightforward manner, these bats are using the Doppler shift.

A greater horseshoe bat shows the horseshoe-shaped nose leaf which acts as a megaphone to concentrate and aim its ultrasonic squeaks.

Below: A serotine with a hawkmoth. Too big to be eaten in flight, the moth has been brought to a perch.

As the bat flies towards an object, uttering a steady stream of clicks, it runs into returning echoes. The echoes have progressively less distance to travel, so they appear to speed up. For every millisecond, the bat meets more echoes, so the effect is of a higher frequency. If the bat is flying away from the object, the distance between echoes is effectively stretched and the frequency appears to drop. Doppler shift is used in modern radar to distinguish moving aircraft from stationary ground objects and constant frequency bats perceive flying insects against a background of foliage or distinguish between two moving insects by the difference in frequency shifts. When horseshoe bats fly low over rough pastures in search of beetles, their long pulses results in a 'mush' of echoes coming back from the grass, but they can pick out moving objects, and large insects are highlighted by the change in echoes caused by their wingbeats.

Efficient as their echolocation is, the interactions of bats and insect prey is not wholly one-sided. Several groups of night-flying moths, including the families Noctuidae and Arctiidae, as well as lacewings, have ears which can pick up the ultrasonic clicks of bats. As the outgoing clicks are many times louder than the echoes, the insect can hear the bat long before the bat becomes aware of it, so there is time for evasive action.

The ears of noctuid moths, which are more strictly and pedantically called hearing organs by biologists because their mechanism is different from the mammalian ear, are tuned to sounds with frequencies between 40 and 80 kHz, which is the range of most bat ultrasonics. When a moth hears a distant bat it starts to behave erratically, looping and spiralling with changes in speed. When the bat comes within 3 metres, and is still unaware of the moth, the reaction is for the moth to plunge earthward. This early warning system is good enough to reduce the chances of being caught from evens to less than one in ten.

The arctiid moths turn the tables on the bats. They produce their own ultrasonics by means of a drum-like organ on the thorax which clicks at the same rate as the wingbeat and produces sounds with frequencies between 30 and 90 kHz. This is well within the range of bats' hearing and serves as a warning that the moths are distasteful in the same way that bright colours warn off day-hunting predators. This was confirmed experimentally by training bats to take mealworms lobbed into the air and then playing recordings of the moths. The bats reacted by swerving away from the mealworms.

Variations of straightforward frequency modulation and constant frequency are employed by the many Microchiropteran bats and the echolocation systems of many have yet to be investigated. The use of echolocation to hunt prey other than flying insects is still something of a mystery. Bats which pick insects off foliage or hunt scorpions, lizards or small rodents often employ quieter clicks; they are sometimes called 'whispering bats'. The two suggestions are that quiet clicks reduce the amount of echoes bouncing from the background or that they reduce the chance of the prey being warned of the bat's approach.

The fishing bats posed a particular problem because sound does not pass easily from air to water. Several bats are known to catch fish, including the European pond bat, but two American bats are fishing specialists. One feeds mainly along the sea shores of Baja California and Mexico and the other is widely distributed over much of Central and South America. It feeds mainly on freshwater fishes but will also hunt over the sea. The two fishing bats have long, curved claws which they use to scoop fish out of the water. At one time it was thought that they merely trailed a foot in water at intervals in the hope of gaffing a fish, but now it is known that they can detect faint ripples of disturbance which indicate fish just under the surface.

Despite the proved efficacy of echolocation for catching prey and avoiding obstacles, the use of vision by Microchiropteran bats should not be overlooked. The expression 'as blind as a bat' maligns their eyesight. Although small, their eyes may be as good as those of similarly-sized mammals, such as shrews and voles. Many of the insect hunters have a crepuscular rhythm of activity and take much of their prey in the twilight hours, when insect activity is also at its peak. While there is still some light in the sky it is likely that they will be able to pick out large moths and beetles as silhouettes. Blindfolded bats fly more erratically and more slowly as they make their way back to the roosts. Eyesight may be valuable for homing when it is useful to be able to recognise landmarks outside the range of echolocation.

The infamous vampire would seem to be a good candidate for using its eyes but it still seems to shun bright moonlight. It is a 'whispering bat' and its low-energy pulses are sufficient for detecting the large animals on which it feeds. Vampires are unique in being the only blood-sucking parasitic mammals. The three species range from Mexico to Peru and Brazil and are an important pest, not only because their blood-drinking weakens animals but also because they carry and transmit certain lethal diseases.

The usual victims of vampires are cattle and horses, but humans are attacked on exposed flesh, and two species regularly attack birds. It has been suggested that dogs are seldom attacked because their hearing is sensitive to the vampires' ultrasonics. The vampires cruise slowly about 1 metre above the ground in search of victims. They can land and move so lightly that the victim is not awakened. Large mammals are sometimes attacked on the legs by the vampire landing several metres away and scuttling over the ground like a huge spider, with its body held well clear. Using razor-sharp incisor teeth, the vampire cuts open a wound so gently that the victim is not alarmed, then bathes the wound in saliva containing an anticoagulant and sips the welling blood using its grooved tongue. The vampire gorges until it can hardly fly but its system is designed to excrete large quantities of waste fluid very rapidly.

The second major group of bats—the Megachiroptera—rely extensively on vision for finding their way. This group consists of the single family of fruit bats, or flying foxes. They are restricted to the Old World, their fruit, pollen and nectar-eating habits being taken over in a remarkable parallel by some of the leaf-nosed bats in the Americas. The fruit bats have very large eyes with pupils which open to a wide stare and a retina designed for sensitivity to very dim light. The retina is folded so that there are 672,000 rods per square millimetre, over four times more dense than in the human eye. Each nerve fibre is linked to 300–400 rods so that it can be stimulated by very low levels of light, at the expense of acuity. Fruit bats roost in trees and do not encounter pitch darkness, with the exception of 11 species of rousette fruit bats which roost in caves and ancient temples. These bats have a simple form of echolocation consisting of audible clicks, like pebbles rattled together, made with the tongue.

The fruit bats live in the forested regions of the Old World tropics, including

Mediaeval Europe had its legend of the blood-drinking vampire, a supernatural being, often in the form of a bat. Real vampires were later discovered in tropical America.

oceanic islands, where there is a succession of ripening fruit and blossoming flowers (some species are nectar or pollen eaters) to provide them with food the year round. The large numbers of bats can make them a pest when they descend on crops of figs, bananas or mangos. The fruit must be ripe because their diet is almost liquid. Fruits are crushed by the flat teeth and rubbed with a strong tongue against the ridged palate, like a lemon squeezer. The juice extracted is then swallowed along with the soft pulp, and the fibrous matter and seeds are spat out in a pellet. As the fruits are often carried away to be consumed, the bats are helping to disperse the seeds and they contribute to the spread of the species, the trees soliciting, so to speak, the assistance of the bats in dispersing their seeds.

Trees visited by bats typically have fruits on long stalks so that they are easy for the bats to reach, they have a strong odour because the bats use a sense of smell as well as vision for foraging, and they continue to hang in the tree when ripe, rather than fall to the ground.

Above: An epauletted fruit bat
eating a wild fig. The species is
named after the tufts of white hair
on the shoulders of the male.

The bat falcon is one of the few
birds which makes a habit of
preying on bats, catching them as
they stream out of their roosts.

The pollen and nectar-eating Megachiropteran bats—fruit bats is hardly a suitable name—have evolved a similar cooperation with their source of food. They have long snouts and extensible tongues armed with a brush at the tip for picking up nectar and pollen. The plants, which include baobabs and wild bananas, have strong, trumpet flowers which open only at night, so their nectar is denied to birds. They are usually white or purple and they have a strong smell of fermentation to attract the bats. In return for a meal, the bats pollinate the flowers as they move among them.

The bats are the most nocturnal of animals. They are very rarely seen by day and the majority have the system of echolocation which enables them to operate in pitch darkness. Consequently, no other group of animals provides such an opportunity for studying the factors which lead to nocturnal life. They can be compared with the birds which share the ability to fly and have frequently adopted similar diets, but which are mainly diurnal. As we have seen, some birds are, however, nocturnal and directly overlap the bats in their activities. Nocturnal birds can interact with bats either by competing for the same food or by predation, and the two must have come into conflict over the course of their evolution.

The earliest known bat is represented by a fossil from the Eocene period, about 50 million years ago. The world was warmer then and sub-tropical forests reached as far north as Britain. The ancestors of the bats were insectivores, similar to modern shrews, which had already evolved a nocturnal and arboreal life. How the bats grew wings and developed the power of flight is as much a matter of conjecture as it is for birds. The oldest fossil bat is already a modern bat with no primitive features which might give a clue to the evolution of the wings. It could be that the ancestral bats began by leaping from bough to bough, like tree squirrels, and then increased their range by gliding, like flying squirrels. Once airborne and by developing control over their movement, they had the potential to exploit the masses of insects which were flying relatively unhindered at night.

The evidence supports the theory that nocturnal birds, the owls and nightjars, have come into conflict with the bats and that both sides have drawn back to avoid competition by specialising. Overall, the bats are significantly smaller than the owls and nightjars and they concentrate on small flying insects. There is an

A flying fox, one of the fruit bats. It may have to live a nomadic life travelling around the country in search of ripe crops of fruit.

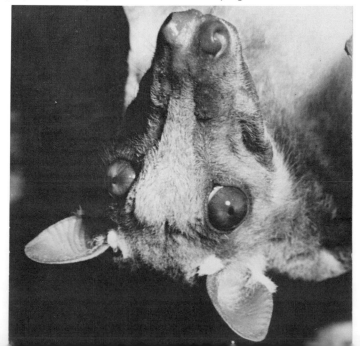

overlap in the size of prey taken by carnivorous bats and small owls but their distributions show that in major life-zones either one or the other dominates. Africa is rich in small owls but has few bats feeding on vertebrates. Tropical America, on the other hand, has several carnivorous bats but very few small owls. There is a similar division in the habit of fishing: fishing bats live in South America and fishing owls in Africa.

Among the fruit and nectar eaters, the bats have the upper hand at night. The only nocturnal fruit-eating bird is the oilbird and the well-developed sense of smell, characteristic of the mammals, must have given the bats an advantage. There must be competition where day-flying birds visit the same trees as the bats but this effect has been reduced where fruit and flowers have evolved a special relationship with bats.

The pressure of competition has, no doubt, restrained the bats from coming out to feed by day, when they would come into conflict with the insect-eating swifts, swallows, flycatchers and bee-eaters, the nectar-drinking hummingbirds and sunbirds, and the fruit-eating toucans and hornbills. Nevertheless, the almost complete absence of day-flying bats is as likely to be due to predation as competition. The night has given bats a high degree of immunity from attack. Although bat remains turn up in owl pellets, owls are not well suited for catching manoeuvrable prey in flight and bats face their greatest hazards at twilight, when day-flying predators are still active.

Two birds of prey have specialised in catching bats and individuals of several species form the habit of preying on them. The bat hawk is a kite which lives in Africa, including Madagascar, and in South-east Asia. It hunts over open spaces, where it quarters the ground in rapid flight and pounces on swallows, swiftlets, large insects and bats. Hunting is confined to half an hour or so around dusk and dawn, but it may hunt deeper into the night when there is a bright moon. Prey must be very abundant if it is to earn a living in such a short time, and bat hawks make a habit of attending the entrances of caves to take advantage of the streams of bats. It does not seem to be able to deal with large fruit bats. The bat falcon of Central and South America has a similar diet. It is not so tied to hunting at dusk but it haunts bat roosts at twilight.

Hunting at twilight is not unusual for birds of prey. European sparrowhawks, kestrels and, in particular, hobbies have been recorded as hunting bats and a bat cave will attract a number of predators. At the Carlsbad Caverns of New Mexico, peregrines and American sparrowhawks (a falcon) make catches from the issuing stream of bats with considerable efficiency, whereas the great horned owl missed more than it caught. It struck at random and the bats usually managed to out-manoeuvre it. The owl was successful only because the bats were flying in a tight formation which gave the owl the chance of a lucky strike.

The relationship between birds and bats is demonstrated by one of the few 'natural experiments' which demonstrate the value of nocturnal behaviour. The Azores and Madeira are inhabited by the European Leisler's bat. On the Azores, it is regularly seen hunting during the day, especially in the afternoon, but on Madeira it is wholly nocturnal. The Azores, set far into the Atlantic, have few birds and no birds of prey, but the bats on Madeira must compete with swifts for aerial insects and run the gauntlet of kestrels and sparrowhawks.

Sunset on the savannah. A small
herd of impala keep alert during
this dangerous period.

The African Savannah

Dawn and dusk are brief in the tropics. There is no extended twilight or gloaming when the country glows in the gentle light from a sun already below the horizon, and there is no magic slow brightening of the sky from darkness to sunrise. The transformation is sudden, but spectacular nonetheless, and there is nowhere better to witness this than on the open grasslands of Africa where the flat country gives an immense panorama perhaps broken by low hills and dotted with trees.

During the short transition from day to night, the sky changes colour rapidly and shades of pink are thrown onto distant mountains and the underside of clouds. Sunrise is preceded by a brief golden glow and is seen to best advantage from the shores of a lake where the still water surface is a mirror reflecting the light and silhouetting flocks of flamingos and pelicans. At sunset, a watcher with good colour vision may be rewarded with seeing the 'green flash', a pillar of pale green light which appears for a fraction of a second on the horizon where the sun has set. The colours of the sunset are often intensified by airborne dust which has been whirled up by wind from the parched countryside or by fine ashes from the frequent bush fires.

Africa conjures up images of the grasslands with scattered flat-topped thorn trees and the park-like open woodland or bush which are collectively known as savannah. The climate of the savannah alternates between long dry spells and shorter rainy seasons which bring an immediate flush of plant growth and an attendant swarm of insect life to the previously parched scene. In certain places, notably along the banks of the rivers and lakes, there is denser forest, but the predominance of short grasses and herbs on the savannah is maintained by fire and by browsing animals which prevent trees from taking over. The savannah lands cover a vast tract of land stretching across Africa. This was the Africa of the big game hunter and explorer and it is now the Africa of the tourist and scientist. It is the home of a mass and diversity of large animals, unrivalled in any part of the world. There are herds of wildebeests, elands and gazelles, as well as many other antelopes, elephants, zebras and buffaloes. They are preyed upon by the big cats—lion, leopard and cheetah, and by hyaenas and wild dogs. Smaller animals, which often go unnoticed, include many rodents, which may have as much impact on the plants as the larger grazers and browsers, and small hunters like the jackals and mongooses. The bird life is also rich and varied, from stately, stalking ostriches to immense flocks of the sparrow-like weaver birds. The savannahs are a wildlife paradise but human settlement has encroached on the once seemingly endless spaces and scenes approaching the original natural state are now found almost wholly in the national parks and reserves.

The parks are open to the public and in some the animals have become used to motor vehicles, which provide a safe and comfortable grandstand view of them in as near as possible natural conditions. However, the animals are often disappointingly inactive and many will be hiding in thickets, sheltered from the hot sun as well as from the gaze of the curious. The best time to see these animals is in the early morning or on moonlit nights, but travel at night is not normally permitted by the park authorities, partly for the visitors' safety but also to give the animals some peace. The explorers and hunters who first penetrated these lands also kept to their tents at night where they were safe from the many predators which were abroad in the dark. They were aware of the nocturnal activity beyond the confines of the camp only through glimpses of shadowy forms with gleaming eyes reflecting the lamplight and by the many calls coming out of the darkness.

No tale of the savannah would be complete without roaring lions, either close enough to the camp to be uncomfortable or as distant rumbles like thunder. Providing there is no immediate danger, it is the most exciting sound to be heard in the whole of the savannah region. The night silence is broken first by one or two preliminary moans, then full-throated roars erupt in a series lasting for half a minute or more, before dying away in a series of hoarse grunts. Roars can be heard over distances of two or three miles and several lions may join in a chorus, calling and answering from all points of the compass. The roar is a simple sound beacon, saying 'Here I am'. It keeps members of a pride in touch, it warns rivals of each other's presence and it can help to intimidate an opponent. Roaring is nearly always heard at night, which is a good indication that the lion is a nocturnal animal; the indolence of lions during the day would seem to bear this out.

Using the familiar fireside cat as a yardstick, it is easy to assume that all cats hunt, and court, at night. Does the lion fit this view? Lions are immediately marked out as being different from all other cats by their communal life. No other cat lives so sociably as a pride of lions. The pride is a cohesive unit based on several adult, related lionesses with their cubs and adolescents. One or more male lions associate with the pride and father the cubs but they take little part in the communal hunts which are the unique part of lion life. According to George Schaller, a male lion looks like 'a moving haystack'; its magnificent mane is a positive handicap when stalking. Most kills are made by the lionesses, which frequently hunt together, but the large and powerful males then move in, drive the females away and literally take the lion's share. Apart from fathering the cubs, the male lions' role is to defend the pride's territory so that the lionesses may hunt and rear their cubs without interference from other lions.

George Schaller spent three years studying the lions of the Serengeti National Park in Tanzania and has made a detailed survey of their daily regime. He found that lions were opportunists and never missed the chance of a meal, when, for instance, they stumbled over a crouching antelope fawn. Nevertheless, they were generally lazy and spent most of their time resting. The lions lying in the shade

of an acacia tree were not merely resting from the night's hunting; they were likely to rest for most of the night as well. As many as 20 or 21 hours out of 24 are spent inactive. The few hours of activity were usually taken at night, particularly in the first hours of darkness. The time of day is not important in itself; the lions wisely choose the conditions when hunting stands the best chance of success.

Its main prey, the zebras, buffaloes and antelopes, can easily outrun a lion. Given a head start, they will show the lion a clean pair of heels. Furthermore, the prey animals realise this and they continue to graze while keeping a watchful eye on an approaching lion. The lion also realises its limitations and does not attempt to chase alert animals. Its success at hunting depends on approaching to within 30 metres before leaping forward in an explosive dash and catching the victim unaware. The lion uses all the skill and patience of a cat stalking a sparrow, and concealment is essential when stalking animals which have good all-round vision and are constantly alert for signs of movement. For this reason alone, lions prefer to hunt by night. Schaller saw lions watching potential prey during the afternoon but waiting until nightfall before starting to hunt them. Even a bright moon will put lions off hunting, until it is obscured by a dark cloud.

Hunting by night was particularly important for Schaller's Serengeti lions because the grassy plains are open and flat, offering little cover. Daylight hunts are more common in the vicinity of woods or thickets where there is cover for the lions to hide in. Even so, the lion has to keep its prey in view and the slightest movement, as it raises its head above the grass or peers through the foliage, may be detected by the prey.

The second largest of the African cats, the beautifully patterned leopard, is another hunter which prefers the cover of darkness. Unlike the social lion, the leopard is solitary and its habits are not so well known as those of the lion because of its secretive behaviour and its preference for hiding up in woodlands. Leopards

A pride of lions disposing of a waterbuck. Most of their kills are made at night, when it is easier for them to stalk unobserved by their victims.

Opposite top: A leopard relaxes in solitude after the night's hunting.

Opposite bottom: The spotted hyaenas' hunting skill was not appreciated until they were kept under observation at night.

take care to keep out of sight but their position is often given away by a tail dangling from a tree. As fits the solitary lifestyle, leopards prey on smaller antelopes and other animals which are easy for a single predator to subdue. They hunt with an almost snakelike stalk, belly to the ground and with a fluid movement. Yet, despite making the best use of cover and taking every precaution to avoid being seen, a leopard's hunting is still most successful when assisted by the cover of darkness.

By contrast, the third big cat of the savannah, the cheetah, is definitely diurnal, although it will hunt on moonlit nights. The cheetah is an odd cat which must have become separated from other cats early in the evolution of the family. It has become specialised for high speed running, in the same way as a greyhound. Surplus weight has been removed and the slender body is long and supple to give the cheetah an enormous bounding stride. It is also the only cat which cannot sheath its claws. They are blunt and leave marks in the ground like a dog's.

The cheetah's hunting ground is the open plains. Like other cats, it has to get within striking distance of its victim and to do this it approaches boldly at first, then stalks slowly, freezing when the victim looks up but not otherwise attempting to hide. Then it breaks into a trot, a stride and finally into a headlong sprint. The victim, usually a gazelle, takes flight and starts to jink as the cheetah closes, but the cheetah can change direction with equal ease until it is near enough to deliver a heavy slap with a paw.

It can only be presumed that this form of hunting, with a prolonged chase and repeated dodging and circling, does not work well at night. One would expect the initial approach would be facilitated by darkness but the cheetah may need to see where it is putting its feet during the fast manoeuvring of the final chase. A trip resulting in a sprained ankle, or worse, could be fatally hampering to a cheetah. It is interesting that cheetahs have a pronounced black 'tear-mark' under the eye and I wonder if this could be an anti-dazzle device like the similar marks on the swift-moving falcons, which also hunt by day.

Two more important savannah carnivores, the wild dog and the spotted hyaena, live in packs. The dog is diurnal but the hyaena is active by day and night. However, it is only in recent years that the nocturnal habits of the hyaena been recognised, and the result has been a complete reappraisal of the species' role in the savannah economy. Until then, the hyaena had been reviled as a cowardly scavenger. There is no doubt that hyaenas are very good at scavenging and do not pass over anything edible. They have powerful jaws armed with both sharp teeth for slicing tough skin and tendons, and stout, flat teeth for crushing bones, so that every scrap of meat, skin and bone can be devoured and absorbed by a very effective digestive system. The notion of the hyaena as a cowardly scavenger comes from a common enough scene on the savannah: a pack of hyaenas, along with a flock of vultures, waiting to clear up the remains of a kill. It was once thought that lions never touched carrion. As Oliver says, in *As you like it*:

> '*Tis the royal disposition of the beast*
> *To prey on nothing that doth seem as dead.*'

That Shakespeare and everyone else were wrong was shown by the Dutch zoologist, Hans Kruuk, when he started to follow hyaenas in a Landrover, trailing their packs at night and observing their habits through binoculars on bright nights. The hyaenas proved to be the hunters and the lions the scavengers.

This is not the first time in the field of natural history that a firmly rooted idea has had to be overthrown when the full facts have become known. The early naturalists and game hunters were often very good observers of wildlife but their deductions were sometimes erroneous. They were mistaken about hyaenas because they had formed an opinion on what they had seen so often by day, and the few reports of hyaenas hunting their own prey were discounted. Only with the systematic approach of the scientist has the record been put straight. Kruuk discovered that spotted hyaenas are extremely good hunters. In the Serengeti National Park, about two-thirds of their food is killed by the hyaenas themselves; the amount varies in other localities depending on how many dead animals there are available for an easier meal obtained by scavenging.

Spotted hyaenas live in packs or clans of up to 100 individuals. They spend much of the day lying up in the shade of trees or rocks or wallowing in puddles, and at nightfall the clan gathers for the night's activities. Led by a senior female, the hyaenas set out to patrol the borders of their territory and search for food.

The border patrol consists of part or all of the clan and it proceeds through the territory and around the boundary, while ignoring opportunities of feeding. Ownership of the territory is demonstrated by leaving scent marks, whose message is reinforced by chasing any strangers found trespassing. Sometimes

94

members of a neighbouring clan chase their prey across the boundary and the home clan rushes across to intercept. An incredible mêlée then breaks out in which hyaenas may be wounded or killed, accompanied by a babel of giggles and yells. Along with the lion's roar and the melancholy whooping of solitary hyaenas, this chorus is one of the evocative sounds of Africa. Harrison Matthews has given a vivid description: 'shrill shrieks and yells, accompanied by deep emetic gurgling and groans, . . . a background for wild peals of maniacal laughter . . . . The hair-raising din is indescribable and is truly horrible to human ears.'

The food of the spotted hyaena includes small animals such as tortoises, hares and flamingos, and large animals such as buffaloes, elephants and rhinoceroses, but the main prey is wildebeests, zebras and gazelles. Hans Kruuk's observations show that the spotted hyaena is far from being a scavenger that sometimes 'jumps the gun', as it were, by killing animals. It is a skilful hunter which uses different tactics for attacking different prey.

Thomson's and Grant's gazelles are small animals and are sometimes hunted by solitary hyaenas at night. The hyaena walks into a herd and picks out one individual which it begins to chase. Gazelle and hyaena set off across country, the gazelle not appearing to be too harried. Only when the hyaena begins to close up does the gazelle begin to run at top speed and to jink, but it is then too late and it is grabbed and bowled over. Strangely, a much larger antelope, the wildebeest, is often also chased by a solitary hyaena. The predator runs among the dim silhouettes of the herd, causing the wildebeests to scatter and regroup. Then it stops and watches the milling animals until it has selected one to chase. Off they go, often for some distance, until other hyaenas eventually join in and the wildebeest is surrounded, brought to bay and torn down.

Attacks on zebras follow a rather different pattern. A single hyaena is no match for a zebra, which makes the best use of its teeth and hooves to fight back. Moreover, zebras cooperate in defence. When danger approaches, family parties stay together. The mares gather up their foals and retreat, with the stallion hanging back to fight a rearguard action. It takes a gang of hyaenas to break through and overwhelm the defences, and only when one zebra has been pulled down and submerged under a mass of hyaenas will the others abandon it and make good their escape.

As with a border clash, the night is made hideous with the eldritch chorus of excited hyaenas as they jostle to tear at the quarry. Being members of one clan, there is no fighting and the blood that coats their fur comes from the carcase, not from their own wounds. The meal is soon finished; one hyaena can eat 15 kilograms, or one-third of its own weight, at a sitting and all that is left of a large kill after one hour is a few hyaenas noisily crunching the bare bones. Hans Kruuk has pointed out that this is one reason why the hyaenas' hunting habits had been overlooked. When hyaenas can feed undisturbed, there is nothing left by morning but, when a lion moves in and drives them off a kill, it feeds slowly and the hyaenas have to wait around for hours before they can get back to clear up the remains. As day breaks, the lion is found on a kill with hyaenas gathered about, and the conclusion is drawn that the lion is the killer and the hyaenas are waiting to scavenge.

The question why the spotted hyaenas should hunt mainly at night has not yet been resolved by research. They do not require the cover of darkness to make a stealthy approach; rather they appear to make a point of approaching boldly and scaring their quarry into running. The wild dog which also lives in packs and employs similar hunting techniques is mainly diurnal. It hunts principally just before dusk and just after dawn, but it will sometimes kill during the day or on moonlight nights. This seems to be one of the many cases where we have, as yet, no explanation for observed patterns of activity.

The last of the large savannah predators differs from the preceding animals in many respects. Although it is an aquatic animal and so not seen on the savannahs, the Nile crocodile has an important role because it preys on the savannah animals as they come to drink. Perhaps this role should be described in the past tense because the swarms of crocodiles which once thronged the lakes and rivers of Africa have dwindled in recent years, to extinction in many places. Disturbance to breeding grounds by curious tourists or land development, hinders the crocodiles' ability to recoup the enormous losses inflicted by the leather trade.

Crocodiles are interesting for their own sake as sole survivors of the Archosaurs, the giant reptiles which once dominated the world as dinosaurs, plesiosaurs, ichthyosaurs and pterosaurs. A study of the crocodiles' way of life helps to reconstruct the habits of these ancient animals. One thing which has emerged is that, as far as crocodiles are concerned, 'cold blooded' should not be used as an epithet for behaviour that lacks the finer feelings of concern for others. Unlike the majority of reptiles which lay a clutch of eggs and then abandon them so that the emerging young must fend for themselves, the female crocodile stays by the nest. When the baby crocodiles are ready to hatch they start to call with insistent croaks until their mother opens the nest and lets them escape. They stay with the female for several weeks, following her and clustering around her body like a family of ducklings. If necessary, she carries the hatchlings delicately to water in her mouth. Moreover, the father, the parent who throughout the Animal Kingdom has notoriously little to do with his offspring, will share these duties.

Where crocodiles still abound, they make an easily viewed spectacle through their habit of gathering to bask on sandbanks. The daily regime is to bask in the morning and afternoon, with a retreat to shade in the midday heat, and a return to the water before sunset. After nightfall, they become active and start to look for food.

A crocodile's diet depends on its size. Baby crocodiles eat water insects and graduate to fish as they grow up. Eventually, they become big enough to tackle large mammals. Sometimes a crocodile will lie in wait by a game trail and ambush an unwary victim, killing it with a blow from its tail or a crushing side-swipe of its head. Alternatively, the crocodile waits in the water where animals come down to drink and, on spotting a victim, submerges and swims underwater to erupt at its target, seizing it underwater. The prey is at a complete disadvantage in the water; all the crocodile has to do is hold on until the prey's struggles subside and it drowns.

Hoofed animals, in fact, make up the minority of the crocodile's victims, although one good kill is worth a score of small fry. Rodents may make up half the prey of even large crocodiles, and the water-loving cane rat is a frequent

After a day spent basking in the sun, the Nile crocodile retires to the water to lie in wait for animals coming down to drink.

victim. The most commonly taken large animals are antelopes such as waterbuck, sitatunga and reedbuck, all of which live near water and are consequently at risk. The hippopotamus is, however, not at risk. Hugh Cott has watched hippos walking among basking crocodiles and if a crocodile did not get out of the way voluntarily, the hippo knocked it clear with its immense snout. Hippo calves may be killed if they stray from mother's side but she can crunch a crocodile clean in two if it ventures too close.

Night is, then, the time when the majority of the large hunting animals are abroad, stalking their prey or lying in ambush. But even for these species, the rule is not inviolate and they will all hunt by day as the occasion and need arise. The question is why they should so often choose to hunt in the dark. As a general rule, hunting animals have eyes which function efficiently in both bright and dim light. They are also assisted by their senses of smell and hearing so that they are capable of operating by day or night. The answer seems to be that the cover of darkness weighs the balance of predator and prey in favour of predators; ambushes and stalking are that much easier when darkness renders the keen eyesight of prey ineffective and the animals vulnerable. The best strategy for prey animals would, therefore, be to make themselves scarce at night, but this is usually impossible. If it was possible, the predators would not hunt at night.

Few of the larger savannah animals can escape to safety at night. Baboons spend the day foraging on the grasslands and rely on numbers for safety. They live in troops of up to 200 individuals under the care of the older males who keep a special look-out for danger. If a predator approaches, the baboons bunch together with the young adult males forming a protective screen. A solitary leopard can be driven off by threat of the baboons' slashing canine teeth but the appearance of a lion pride causes a retreat to safety. At nightfall, the troop retires to lofty perches on trees and cliffs. The warthog, an animal so ugly as to be appealing to the human eye, also has a safe retreat. Like the baboons it is diurnal and spends the night in an underground burrow, although it also will come out in the moonlight.

In contrast to the baboon and the warthog, most of the hoofed animals on the savannah are active, at intervals, throughout the day and night. The large flesh-eaters can gorge themselves on a good kill and will not hunt again for a day at least but the grazers and browsers need frequent meals of their bulky, nutrient-poor plant food. Furthermore, the process of feeeding takes up a fair proportion of the 24 hours.

The elephant, largest of the savannah animals, and so little concerned with avoiding predators, is active at intervals through the day. It spends three-quarters of its time feeding, ripping up and chewing large quantities of foliage, grass and fruit. The main meals are taken in the morning, the afternoon and around midnight. The midday heat is avoided by taking a light doze in the shade but the main sleep takes place in the early hours of the morning. An elephant will doze on its feet like a horse but, showing its indifference to predators, during its main sleep it lies down and snores. Such a pattern of activity classes the elephant as more of a diurnal animal but it is capable of nocturnal activity and, where disturbed by human presence, it shifts to a more nocturnal regime. It has poor eyesight and probably relies largely on its trunk to find its way about.

Medium or large antelopes such as the impala also have a bias towards activity during the day. Impala can be found feeding at any time but herds synchronise their activities and feed in bouts which mainly occur in the day. There is usually an intensive period of grazing, grooming and territorial sparring by the males in the late afternoon which ceases abruptly at sunset. Then the antelopes lie down to ruminate. Another meal is taken at midnight, followed by more rumination until dawn when they get up, groom for a few minutes and start feeding again.

The impala and similar antelopes live in grasslands with scattered trees. Their main line of defence is their eyes, which are constantly searching for predators in the sparse cover, and their habit of forming herds increases the chances of detecting danger. If one antelope is scared, it communicates its suspicions to its fellows by looking alert and barking or snorting, so that there is safety in numbers. However, darkness robs them of their main defence and the night is therefore a time for lying quietly to avoid detection. The midnight feed corresponds to a quiet period for the predators which are more active near dusk and dawn.

Smaller antelopes reverse this strategy and become active at dusk, which would seem to be the most risky time. However, such species as dik-diks and bushbucks lead a very different life from impala. They are solitary, living alone or in pairs, and they live in denser vegetation, so they are relying on remaining hidden and undetected rather than on the collective vigilance of a group.

When alerted, their reaction is to stand or crouch stock still to avoid being noticed. The same stratagem is used by advancing infantrymen when caught in the light of a flare. If they ran for cover or dropped to the ground, they would be spotted instantly. In dim light, the eye is very perceptive of movement, but will overlook shapes, particularly if the outline is broken up by intervening twigs and stems or by patterns on an animal's coat.

The function of the zebra's stripes has been disputed for years, but there is a good case for their being for concealment. While zebras are very conspicuous in broad daylight, they become indistinct and merge into the background as the

light dims. The same disruptive striping has been used in the past on warships and tanks. Their outlines become hard to pick out at long distances or in dim light and range-finding is difficult. As for zebras, experienced travellers in Africa have reported failing to see zebras under starlight when so close that they can be heard breathing, whereas antelopes of a comparable size can be spotted at four times the distance.

The lives of these hoofed animals are not regulated solely by the need to avoid being eaten. In the tropical regions, the day can be a period of stress for animals that have to feed in open terrain. The heat of the sun is not only uncomfortable but the animal has to keep cool by losing water through sweating or panting. If it is living in dry country water loss can be a serious problem and has to be kept to a minimum. One way is to avoid exertion and find some shade, and general lethargy overtakes the animals as the savannah disappears in the distortion of a heat haze. As the sun sinks and the temperature drops, there is a general arousal which for many animals carries well into the night. Providing the air is not too dry, there is a positive advantage in feeding after nightfall. When the air cools, it becomes more humid, and plants, which have wilted during the day, absorb moisture and revive. The animals then benefit from eating the fresh, often dew-soaked herbage.

The best exponent of the art of keeping cool by day and becoming active in the open by night is the hippopotamus. Its name means 'water horse' although 'water pig' would be better. It is well adapted for an aquatic life with its ears, eyes and nostrils set on the top of the head so that it can remain almost submerged. Hippos retire to the water an hour or two before sunrise and, with perhaps an hour's basking out of water at midday, they stay there until a few hours past sunset. The hippos then come ashore and make their way along regular, well-beaten trails in search of grazing. The trails extend as much as 10 kilometres inland and the hippos spread out on each side to feed on areas of grass.

Two hundred kilograms of grass is needed to fill the stomach of a single hippopotamus, so where these animals are common, they have a considerable effect on the vegetation. Unless they are overabundant and overgrazing leads to destruction of the plant cover and erosion of the soil, the presence of hippos can be beneficial to other grazers by maintaining pastures of open grassland. The masses of grass they remove are carried back to the water in their stomachs and intestines and eventually their faeces fertilise it so that hippopotamuses contribute to the dense populations of fish and the large numbers of water birds.

Details of the hippos' nocturnal wanderings are scant; they have not been subjected to the intensive study that other species have received. They live mostly in groups during the day sometimes tightly packed with heads resting on each other's backs, but spread out at night. Mothers have their youngsters following closely at heel (they are chastised if they stray) but otherwise grazing seems to be an individual business. When on land, hippos spread their dung by flicking it with a windscreen-wiper flagging of the tail. The resultant smelly mess scattered over the ground and bushes is an unmistakable olfactory beacon. Its function may be to mark out territories or ownership of the trails and could be used as a navigational aid to guide the hippos to and from their feeding grounds.

A healthy hippopotamus is a match for any animal. Its treatment of crocodiles could be described as contemptuous and, if threatened on land, it can dash back to the water at speeds of up to 50 kph (30 mph), although this is rarely necessary.

100

Hippopotamuses fear no other animal but they prefer to leave their wallows at night as the hot sun makes them uncomfortable.

Hippos walk through prides of lions scattering them as easily as crocodiles, so they cannot be nocturnal for reasons of safety.

The hippopotamus' lifestyle poses some problems. It is odd that it should be an aquatic animal when it derives neither protection from enemies nor food from the water. The most likely reason that the hippopotamus spends the day in water is because it is susceptible to overheating. It is significant that hippos are usually seen abroad by day only on cool, overcast days. The skin is thick, up to 35 mm on the flanks, but it is hairless and presumably hippos would suffer in the full heat of the sun if they grazed by day. The skin is also rather remarkable because it is not very watertight; water passes through it considerably faster than through human skin. Why this should be so is a matter for conjecture. One suggestion is that the water evaporates and keeps the hippo cool, like the water 'sweating' from a canvas water bucket, but this has yet to be proven.

There was once a story that hippos sweated blood. The skin is provided with glands which secrete, not sweat but a pinkish viscous fluid which dries on the skin. Its function has not been fully explained but it is known to be opaque to ultraviolet rays, which cause sunburn. If hippo 'sweat' is painted on human skin, the area covered remains pale while surrounding parts become burnt. It is possible that the 'sweat' is the original suntan lotion and protects the hippo's skin when it is basking and when it is wallowing in water too shallow to absorb the sun's ultraviolet.

101

While these questions remain to be answered, the indication is that the amphibious and nocturnal lifestyle of the hippo is a means of keeping cool. Somewhere in its evolutionary history it has taken the turning to its present lifestyle. The other three African pachyderms—elephant, and the two rhinoceroses—also have thick, hairless skins, but they have not solved the problem of overheating by becoming amphibious.

Once abundant on the savannahs the two rhinoceroses are now reduced to a pitiful fraction of their former numbers. The black rhinoceros is a browser, feeding on leaves and twigs while the white rhinoceros is a grazer and the difference in diet is reflected in the shape of the mouth. The upper lip of the black rhinoceros is pointed and used for plucking foliage, whilst that of the white rhinoceros is square for cropping grass. The distinction can be remembered by the fact that the white rhinoceros gets its name not from its colour but from the Afrikaans word *wijd* for wide. Although neither rhinoceros is strictly nocturnal, both are active mainly by night and prefer to spend the day resting in the shade or wallowing in mud. The fourth large, hairless mammal, the elephant, would seem to set at nought the theory that hippos and rhinos feed at night because it is cooler. Elephants tend to be more active by day but their big ears, aptly described as looking like maps of Africa, act as radiators, cooling the blood as it flows through them.

Lions, leopards and hyaenas, antelopes, zebra, elephants and hippos are the familiar animals of the African savannah and are the main attraction of the national park. However, there are many kinds of lesser animals which are usually overlooked by visitors. Apart from being less spectacular and well-known, they are more difficult to spot and many are nocturnal.

At any time of a day or night there will be a host of rodents feeding among the grasses or climbing in the trees and bushes. Some show a distinct preference for a particular time of day. Squirrels are diurnal, as they are in other parts of the world. The bush squirrel lives in trees but ground squirrels live in open grassland. They stand upright on hindlegs while feeding to keep a watch and avoid being ambushed, as does the diurnal grass rat. Multimammate rats, so-called after the two rows of up to 12 teats a side, and the pigmy mouse, the smallest rodent in the world, are nocturnal but the rodent which is likely to be seen at night is the springhare or springhaas. This is a rather odd animal that zoologists have at various times classed with the squirrels, the porcupines and the jerboas. It is now placed in a family of its own.

The springhare is present in all the drier savannah lands south of Kenya. About the size of a rabbit, it is unusually large for a rodent and has long hindlegs, short front legs, a long bushy tail and a hopping gait. The forefeet are used to dig the extensive burrows in which it lives during the day, emerging only at night. The presence of springhares on the savannah at night is given away by pairs of red dots of light which bob slowly past in car headlights.

Most of the rodents are vegetarian but a few take advantage of the abundance of insects and other invertebrate animals which come out at night. Insects, such as moths, ants and termites, and the spiders and scorpions which prey upon them, as well as soft-bodied slugs and snails, and many-legged centipedes and millipedes attract a variety of nocturnal animals, large and small. Even lions have been known to eat insects. Many small animals are specialist insect-eaters.

102

A pair of bobbing dots of red light give away the presence of the springhare which looks and behaves like a small kangaroo.

The true insectivores are a group of mammals which includes the hedgehogs and shrews, whose African members differ little from more familiar species, but there are some less familiar insect-eaters.

The aardvark, whose name is Afrikaans for 'earth pig', is an odd-looking animal. Its stout body is covered with sparse hair; its long head bears a pig-like snout and long donkey-like ears and its sturdy feet are armed with strong claws. The aardvark is a skilled excavator; not only does it dig the burrow where it spends the day, when threatened in the open it will quickly dig another. If this is not possible, it defends itself with its claws and, if seized, it turns a somersault to throw its adversary, kicking with its hindlegs at the same time.

Aardvarks are found throughout the less wooded areas of Africa south of the Sahara but its distribution is tied to the presence of its main food, the mound-nesting termites. They are equipped with a 40 centimetre sticky tongue which readily wipes up the teeming insects and withdraws them into the mouth. The nostrils can be shut and are fringed with stiff bristles to exclude the insects while feeding.

Its nocturnal habit and sparse distribution has made the aardvark a difficult animal to observe. Apart from chance encounters, there have been few opportunities for watching aardvark habits. It emerges from its burrow with great caution, waiting for some minutes with only the head showing to test the night air for sound or scent of danger. Once satisfied that the coast is clear, the aardvark sets off, zigzagging over the ground to cover a track about 30 metres across. It is constantly alert, its mobile ears scanning for faint rustlings of insect hordes, and it stops at intervals to sniff noisily at the ground.

Sometimes the aardvark finds termites or ants moving on the surface, especially in damp weather, otherwise it has to dig out their nests. Termite mounds are veritable fortresses of baked earth mixed with termite saliva and excrement but the aardvark can tear into them with its claws. Small nests may be destroyed completely but large termite mounds are 'tapped' by sinking a 30–40 centimetre hole in the side. The aardvark inserts its snout and takes its fill of termites by probing the nest galleries with its sticky tongue. After the aardvark has gone, the termites repair the breach and maintain the ingenious air-conditioning system which keeps the nest cool, moist and ventilated.

The aardvark posed a problem for zoologists who had difficulty in deciding its relationship with other mammals, and it is now placed in a family of its own. A similar solution has been proposed for a second termite-eater, the aardwolf. This looks like a small hyaena but instead of strong jaws and bone-shattering dentition, it has weak jaws and small cheek teeth which are unsuitable for tough

103

food. The aardwolf is often placed in its own family as no more than a distant
relative of the true hyaenas. Its diet is composed almost entirely of termites with
some other insects and spiders, but unlike the aardvark, it is unable to break open
termite mounds and so it has to gather insects from the ground, using a long
sticky tongue.

The aardwolf's day is spent in a rock crevice or an abandoned aardvark hole,
or it may dig its own extensive burrow. Like the aardvark, it leaves its hiding
place at night and quarters the ground in search of food. It walks with its head
down and ears constantly turning until it locates a swarm of insects, and then sets
to work with astonishing speed. One dead aardwolf was found to have 40,000
termites in its stomach and it was estimated that they must have been consumed
within the space of three hours. The aardwolf has to snatch each meal rapidly
because the disturbed termite workers rush back to the nest, while soldier
termites rush out to thwart the enemy with squirts of sticky fluid.

In the economy of the savannah, there is an intermediate stage between the
small plant- and insect-eaters and the large flesh-eaters, which is occupied by
small carnivores—the mongooses, weasels, cats and jackals. Whilst efficient and
well-armed predators, their small size makes them vulnerable to attack by larger
animals and they have to be on the alert as much as the vegetarians. Nocturnal
behaviour is one form of defence available to these animals but this way of life
may be adopted to take advantage of prey animals which are themselves
nocturnal. Nevertheless, there is little doubt that the attentions of larger
predators have driven some smaller carnivores to seek protection from darkness.
The mongooses are vulnerable to attacks from eagles and five of the eleven
African mongooses are nocturnal. The remainder are diurnal and it can be no
coincidence that they are social and go foraging in groups. Like herds of impala,
many pairs of eyes keep a better watch for danger and many sets of teeth offer a
greater threat to would-be attackers. When a meerkat, an appealing mongoose
with the habit of sitting up on its haunches, sees a hawk or eagle, it barks the
alarm to its fellows and they watch the bird closely until it has flown past.

The one exception to the rule of nocturnal and solitary or diurnal and
gregarious is provided by the slender mongoose, which is diurnal and solitary. It
is found from light wooded plains to forests, but its arboreal habits probably save
the integrity of the correlation. The slender mongoose is an agile climber and,
when alarmed on the ground, it heads for safety among the branches at top
speed. The Egyptian mongoose is active by day or night and it sometimes forages
in small parties but there does not appear to be any information that would link
its communal life with diurnal habit.

Some mongooses defend themselves by squirting an evil-smelling fluid in the
manner of a skunk. The African striped weasel and the zorilla share this trait
and, like skunks, they have conspicuous black and white coloration which serves
as a warning that it is unwise to molest them. A larger relative, the ratel or honey
badger is white on the top of the head and brown or black elsewhere as a sign that
it is a very stern adversary. Strong claws and teeth are used in attack, a foul smell
is given off in defence and a tough, loose skin forms an armour against bites from
dogs or snakes. Ratels seem to be quite fearless. They attack animals as large as
buffaloes if they feel threatened. Even so they are basically nocturnal, probably
because this is the best feeding time.

104

Tropical Forests

Now disappearing fast under attacks by fire, axe and saw, tropical forests have the most prolific wildlife in the world. Each forest has a dazzling array of animals and plants, many of which have still to be catalogued. When compared with other places, tropical forests are characterised by a relatively small number of individuals of a very large number of species. Under intense competition, their adaptations for specialised ways of life, for securing food, avoiding enemies and reproducing their kind, have run riot. As each continent, except Europe and Antarctica, has a unique fauna in its tropical forests, a general discussion of their nocturnal life is complicated, although there are many parallels between the continents.

The interwoven webs of competition and predation formed by this variety extended in four dimensions. The many life-forms are spaced out not only through the structure of the forests, but in time. As the sun sets or rises, one collection of animals departs to rest and a second shift takes over. The rapidity of nightfall, with hardly any twilight intervening, that is such a feature of tropical life is modified in the forests. The dense roof of the canopy and a frequent cover of cloud cuts out so much light that it is perpetually dim at ground level. The range of temperature change is also damped down and humidity remains high throughout the 24 hours, so that tender, sensitive animals—millipedes, leeches, flatworms, slugs and snails—are under less constraint to hide by day, but night is the time to run acrosss the less pleasant animals: large, hairy spiders, scorpions and centipedes. The nocturnal insects come out in flocks and swarm around lights and all these small animals attract the nocturnal predators: praying mantises, geckos, snakes, bats and small carnivorous mammals.

For the human visitor to the forests, safely encamped for the night, the changeover from day to night shift is marked by a change in animal calls. By day, the calls of birds seem only to accentuate the quiet gloom—a cathedral-like stillness is the popular phrase—and as the birds retire to roost they are replaced by choruses of frogs. There will have been frogs calling during the day, but the majority strike up at night, croaking, chirping, tonking and boinking continuously and, eventually, monotonously boringly.

Many of the tropical forest frogs have taken to living in the trees, for there is little standing water on the forest floor. They have large eyes which may be partly

106

for seeing well at night, although the frog eye is otherwise adapted for vision throughout the 24 hours. Good vision may be more a necessity for judging distances when jumping. Several families of frogs have taken up arboreal life and all have grown an extra cartilage in each finger and toe. This makes the terminal bone very mobile and thus able to stay pressed against a leaf or stem while the fingers and toes move. The tip of each digit is disc shaped and fitted with a pad which secretes a sticky mucus to give a better grip.

The noise made by the frogs is compounded by the din produced by crickets and cicadas. The frogs' calls are vocal and are generated by the passage of air across vocal cords, but crickets' and cicadas' notes are instrumental. The crickets sing by rubbing a toothed ridge on the underside of one front wing against the edge of the other to set them vibrating. The cicadas, which are sap-sucking bugs like garden aphids, set up their intense, penetrating whine with a pair of sound-producing organs on the abdomen. Each consists of a membrane called a tymbal, stretched inside a stiff ring like the skin of a drum. The tymbal is pulled in and released by a muscle so that it pops in and out like a biscuit tin lid being distorted. The tymbals are set into oscillation at a rate of 100–500 times per second and the quality of the sound is modified by muscles attached to the side of the membrane.

The shrill songs of male cicadas make a deafening chorus in warm countries.

The songs of frogs, crickets and cicadas are used by the males for attracting the females and each species produces its unique signature tune by which it can be recognised as readily as an ornithologist identifies a bird by its call. Sound is a

good medium for communication not only at night but also in the dense foliage
of forests where leaves impede the use of vision. To contradict this reasonable
generalisation, however, a spectacular visual show is put on by the fireflies.

Often called lightning bugs in North America, fireflies belong to two families
of beetles which are found mainly in the tropics although some live in the
temperate regions, for example the European glow-worm. The light given off by
fireflies is not very powerful but it appears bright because it is composed of
wavelengths to which the human eye is most sensitive. In some parts of the world
people have used fireflies as a cheap form of lighting. In Brazil, for instance, they
were kept in lanterns made from finely perforated gourds, or were worn in the
hair or tied to the ankles to give light when out walking at night.

The yellowish-green light is produced by a substance called luciferin. This is
stored in an organ shaped like a motorcar headlight, with a transparent window
of cuticle and backed by a dense layer which acts as a reflector. The luciferin
emits light when it is oxidised by air from the firefly's respiratory system in the
presence of an enzyme called luciferase, and oxyluciferin is formed. Later the
oxyluciferin is changed back to luciferin and the process is repeated. A firefly's
glow is an eerie, cold light and, unlike the light from an electric light, it is,
indeed, cold. The electric light bulb gives out only 10 per cent of its energy as
light, the rest is lost as heat. For luciferin, the figure is 95 per cent for light
emission and only 5 per cent for heat.

The light organs of fireflies are borne on the underside of the abdomen and they are usually present in both sexes. In some species, the female has the stronger light and in others it is the male, but it is always the male which seeks the female. She may be wingless, resembling a grub in appearance, hence the name glow-worm for some species. The 'night-train' or 'railway-worm' is a South American species which has a row of greenish light down each side of the body and a red light on the head. The few diurnal fireflies lack lights.

Wingless stationary females flash their lights to attract males and they switch off permanently after mating. Where the male is brightly lit, he signals to the waiting females who signal in return. In both cases the pattern of flashes is used for identification of the species, and it is possible to attract fireflies by mimicking them with an electric torch. Female *Photuris* fireflies use this method to attract prey. By mimicking the flash patterns of the females of other species, they act as miniature illuminated sirens luring the males of these species into their clutches and eating them. Each species of *Photuris* can mimic several species of dupe fireflies.

The most spectacular displays of fireflies are found in tropical Asia, where male fireflies gather to flash in their thousands or even millions. They choose favourite trees where they foregather every night and flash in unison, with waves of flashing travelling vertically or horizontally through the swarm.

One would have thought that the fireflies' beacons would serve as a great attraction to predators but in fact the insects are ignored. Their bodies contain toxic substances and it seems that predatory animals soon learn to associate flashing lights with an unpleasant mouthful and some fireflies are known to start flashing a warning when handled. The poisons are called bufodienolides (from the Latin *bufo*—toad) because they were first located in the bodies of toads and frogs. This may explain why frogs have been found so gorged with fireflies that they are lit up from within yet show no signs of ill-effect.

The great mystery of the fireflies is why their eggs, larvae and pupae should also glow. Suggestions have been made that this is defensive—startling the predator or warning it that such objects are distasteful. Another suggestion is that the larvae use the glow for luring prey. Many of the larvae are predators but as they eat snails and slugs, luring a meal could result in a long wait!

Less well known over most of the world, but familiar as a tourist attraction in New Zealand, is another spectacularly glowing insect. The glow-worm fly is a form of gnat which lives mostly in humid forests, but also on grasslands and in caves. Both sexes have light organs and the female flashes to attract her mate, sometimes while still in the pupal case. The larvae also glow, but there is no mystery as with the firefly grubs. The glow-worm fly larvae use their light to lure prey by a seemingly ingenious method. Each larva lives in a silk tube dangling a number of strings, studded with sticky droplets. It lies in its tube with the glowing abdomen shining down, making the droplets glisten and act as lures for insects. When one becomes caught and entangled, the larva leans out and hauls up the line.

The smaller forms of nocturnal life, mosquitoes, spiders and scorpions, are not generally welcome in human dwellings but geckos are often treated almost as pets. These small lizards are found in all tropical regions. Some live in

A gecko with a flying termite caught during one of their nocturnal swarms.

rocky or sandy places but the majority are tree dwellers and these species commonly come into houses, especially where there is a thatched roof to live in and lights to attract insects. Geckos are the only reptiles to become truly vocal. The males utter hard, chattering calls from which the name gecko and specific names, such as the tokay and chikchak of Asia, are derived. Some gecko calls, including that of the tokay are so loud as to detract from the pleasure of having these attractive little reptiles in the house but they feed entirely on insects and they endear themselves with their miraculous ability to run up and down walls and across ceilings as if gravity did not exist.

The geckos' secret lies in the fine structure of the toes. The underside of each toe is soled with rows of pads covered with a dense pile of microscopic bristles. The top of each toe is furnished with a flat disc which acts as a suction pad or a hook to grip smooth or rough surfaces respectively. The wriggling gait of a gecko is due to the need to disengage its toes after each step. Instead of lifting the heel first, it curls its toes back, starting at the claw.

A few, usually brightly coloured, geckos are diurnal but most are nocturnal. They have large eyes and during the day the pupil contracts to a vertical slit which

111

Kraits are responsible for thousands of human fatalities every year. They lie on paths at night and strike at those who unwarily pass too close.

is so narrow that it is virtually closed except where small notches on each side meet up to make four pinholes for the gecko to peer through without being blinded while it indulges in sunbathing. Geckos cannot close their eyes in the accepted sense. Their eyelids are permanently closed but the lower lid is transparent to form a window and the alternative protection for the retina is needed when the gecko basks in the sun.

Geckos are harmless, like nearly all lizards, although there are widespread beliefs that their bites are dangerous. Fears of the night in tropical forests are better reserved for the nocturnal snakes which pose a real threat. It would be true to say that snakes are as afraid of people as the other way round and they prefer to slip away when disturbed, but they defend themselves if cornered or surprised. The kraits of Asia prey mainly on other snakes but they account for 15,000 human fatalities a year in India. The venom is extremely toxic and a krait strikes out when provoked by a bare foot treading on it or by a sleeper turning over and trapping it while it is foraging through a dwelling.

In South America, the fer de lance is much feared for its attacks on man. There are two stories of how the fer de lance reached the West Indies from the South American mainland. One is that they were brought over by plantation owners to dissuade their slaves from escaping into the forest, but an earlier account tells of the mainland Arawaks bringing over snakes to terrorise the West Indian Caribs with whom they were at war.

Opposite: The sloth fits its few hours of leisurely activity into the night.

Many snakes have eyes with a vertical slit pupil which guards the sensitive

112

The pygmy anteater lives in the trees, where its prehensile tail and double-jointed back feet give it an extremely firm grip on the branches.

retina when they bask in the sun, and dilates at night, but snakes can rely on three alternative senses to find their prey. Although almost deaf to airborne sounds, snakes are sensitive to vibrations in the ground which are transmitted to the inner ear through the bones of the jaw and skull, but this may have no more value than alerting the snake to a disturbance. The sense of smell is used for recognition of the sexes in courtship as well as for following trails and identifying prey. Snakes have two organs of smell: the nose and Jacobson's organ. The latter consists of two pouches in the roof of the mouth. These are lined with sensory tissue and microscopic particles of odour from the ground or floating in the air are brought to the organ by the flickering, forked tongue.

The third sense available to nocturnal snakes is unique in the animal kingdom. Pit-vipers and their relatives, which include the rattlesnakes, and many of the pythons and boas, make use of the body heat of their victims to direct their attacks. The pit-vipers are named after the deep pits which lie on each side of the face between the nostril and the eye. The pits are very sensitive to heat. They can detect the warmth given off by a human hand held 30 centimetres away and pinpoint its position. Each pit has a 'field of vision' which overlaps slightly with the other so that the pair works stereoscopically in the same way as eyes.

Snakes prey on the wide variety of small animals, including geckos, sleeping birds, even other snakes. The Cuban boa specialises in hunting bats, but small rodents make up the bulk of many snakes' prey. There is a remarkable array of

114

rats and mice living in tropical forests, many of which are active at night and they are joined by some of the more unusual mammals.

South America is the home of the order of mammals known as the edentates, a name meaning without teeth, although some have a few weak cheek teeth. One group, the anteaters, lack teeth entirely and feed almost solely on ants, which are swept up by an extremely long tongue coated with sticky saliva. The giant anteater is an improbable-looking animal, grey with a diagonal, white-bordered, black stripe across the flanks and a bushy tail. The small head is dominated by a long beak-like snout, and the front feet are turned so that the anteater walks on its knuckles with the long claws held off the ground. These claws are used for ripping open ant and termite nests so that the tongue can be inserted, but they are also used for slashing adversaries. Giant anteaters seem to be mainly active by day in areas where they are left alone but they have become nocturnal in settled places. The two smaller species, the tamandua or collared anteater and the two-toed or silky anteater, are both nocturnal and arboreal. The latter has a prehensile tail and 'double-jointed' rear feet which enable it to grasp branches.

Related to the anteaters and also living in the forests of tropical America, the sloths are the most bizarre of all mammals. They are active by night, if such a phrase can be used for animals which are proverbially sluggish and doze for eighteen hours out of the twenty-four, and they spend much of their time hanging upside down by their hook-like claws. They feed on leaves and fruits in

115

A pangolin shows her acrobatic skill as she hangs by her tail while her baby clings to her back.

this position, and they even give birth upside-down, the baby struggling onto its mother's chest and hooking itself into her hair. The lie of the hair has become adapted to the inverted lifestyle and falls from belly to back, so that rain will still run off. The hairs themselves are corrugated, the grooves harbouring microscopic algae which give the sloth a greenish tinge to complete its concealment among the leaves. A sleeping sloth is reputed to look like a bunch of dead leaves or a wasp's nest as it hangs, hunched up, from a branch.

On the ground, under the arboreal sloths and anteaters, the armadillos, another family of edentates, root about for small animals and, sometimes, plant food. The armour plating of armadillos is made up of horny plaques set in the

The flying lemur is a superb
glider rather than a flier.
Normally nocturnal to avoid the
birds of prey, it will become
airborne in the day when
disturbed.

skin and forming rigid plates linked by flexible bands of skin to give the animal
flexibility undreamed of by mediaeval knights. Armadillos are more typical of
open grassland—the savannahs and pampas of South America—but some live
in the forests. By and large they are nocturnal although they can sometimes be
seen out by day. It could be that their small prey is more active by night but this
is not known for certain. As far as the habits of its fauna are concerned, South
America is the Dark Continent. It has not received the same attention from
naturalists as have Africa and Asia.

The Old World cannot match the New in strange forms of forest mammals,
except with the pangolins. Seven species of this order live in the tropical forests
and open country of Africa and South-east Asia. They resemble the anteaters in
general body form and habits and were once classed with the edentates. The
alternative name of scaly anteaters is well deserved. The pangolins have the long
slender snout and strongly clawed feet of the anteaters and are covered with an
armour of overlapping scales like roof tiles, giving them the appearance of an
animated fir-cone. Three of the African pangolins live in rain-forests, while the
fourth lives on savannahs. Two of the forest species, the long-tailed and
small-scaled pangolins, are tree-dwellers; the remainder live on the ground,
although they can climb well. The tree-dwellers can hang by their prehensile
tails and sleep rolled in a ball lodged in the fork of a tree. The ground-dwellers
sleep in burrows, also rolled up. The name pangolin comes from the Malay

117

peng-goling—the roller. It is very difficult to unroll a pangolin and only large, determined predators can deal with them.

All pangolins are nocturnal and all feed on ants and termites gathered with their long, sticky tongues, breaking open nests if necessary with their strong claws. The specialised ant and termite eaters have many features in common. The American anteaters, the Old World pangolins and the aardvark of Africa, which is not a forest dweller, have strong claws for excavating nests, long snouts and long sticky tongues. The teeth are missing or few in number. The aardwolf, an aberrant hyaena living in the same region as the aardvark, has reduced teeth and a sticky tongue but lacks the other ant-eater characters. In the forests of Australia and New Guinea ant-eating mammals are represented by the echidnas or spiny anteaters. These egg-laying relatives of the platypus live in a wide variety of habitats, including tropical forests, feeding on ants and termites. They have strong claws, an extremely slender snout and a very long sticky tongue.

Most of these ant-eaters are nocturnal. The giant anteater is diurnal; the echidnas were once thought to be nocturnal but they are often active by day and the marsupial numbat or banded anteater of Australian eucalyptus forests is diurnal. Without these exceptions we could hazard a guess that ants and termites are easier to obtain at night but too little is known of these animals and their relations with their prey for speculation to be sensible.

Speculation on another diverse group of animals which have adopted a similar lifestyle is rather more rewarding. The forests of the Old World tropics are inhabited by a number of gliding animals. They are incapable of the true flapping flight of birds and bats and merely glide from tree to tree, suggesting, it is reasonable to assume, the manner in which birds and bats first evolved.

Most of the gliders live in the Orient. There is the rare flying frog with outsize webbed feet which act as wing surfaces to increase the range of its leaps. It also holds the underside of the body concave to increase lift. The twelve species of flying lizard or flying dragon have membranes of skin stretched between extensions of the ribs. At rest, the 'wings' are closed and they are opened, like an umbrella, by the ribs swinging out. The advantage of this system over other flight surfaces, including the wings of bats and birds, is that all the limbs are free for landing and for locomotion. To take off, a flying lizard turns head-down on the trunk of a tree and launches itself into space. As it approaches its destination it brakes and pitches upward to land head-up. Flying lizards are members of the numerous agama family, but there is also a flying gecko, which has flaps of skin around its body, feet and tail. The flying snakes are unexpected members of the gliding community. At first sight, a snake has all the flying characteristics of a length of rope but the flying snakes fling themselves into the air and immediately arch the body and hold it in a rigid position to form a hollow underneath to gain some lift.

Among the mammals, gliding has evolved six times, twice in the rodents, three times in the marsupials and sixthly in the flying lemur, which is in a group of its own. The squirrels, the first of the rodents to be considered, are already predisposed to flight. The ordinary tree squirrels are capable of making long leaps with the feathery tail acting as rudder and balancer, but the flying squirrels have taken this ability further with the aid of a membrane of furry skin stretching between the fore and hind legs. Extra support is given by a spur of cartilage on

118

the wrist and the membrane contains muscle fibres with which the tension in the membrane can be altered. Aided by the tail and trimming the membrane as a sailor trims his sails, a flying squirrel can steer in mid air, even changing its mind and returning to the same tree. Before taking off, the flying squirrel moves its head from side to side to help judge the distance to its destination. If it has to leap in a hurry, it may well miss and come down on the ground. Just before landing, it raises its tail to throw its body upright and make a four-point landing. Glides of over 100 metres have been recorded and there are claims for flights of several thousand metres by flying squirrels using air currents to keep them aloft.

Of the 37 species of flying squirrel, one lives in North America, one in Europe and northern Asia, and the remainder are in southern and South-east Asia. The scaly-tails are their equivalent in Africa. These are rodents living in the rain-forests of West Africa and, of nine species, all except one have a membrane stretched between the legs and there is a stiffening rod of cartilage running back from the elbow.

The gliders and flying phalangers are the marsupials' contribution to the conquest of the air. The four species have evolved independently from the three types of possum, but all are very similar to the flying squirrels and were once classed as such. The flying lemur, or colugo, is classed on its own, in the order Dermoptera, and is probably distantly related to the insectivores. As they are not lemurs, and do not strictly fly, the two species are best called colugos. They live in South-east Asia, from Burma to the Philippines and the flight membrane is extended beyond the limbs to join the chin at one end and the tip of the tail at the other.

There must be a considerable advantage for an arboreal animal to develop the powers of leaping and gliding. Whether to escape predators or to search for food, it saves the lengthy and dangerous descent to the ground. Nevertheless, travel through the air is, itself, hazardous. None of these gliders flies very fast, 15 kph (10 mph) is the maximum, and they are not very manoeuvrable. They would make an easy target for birds of prey, so it is not surprising, indeed it is very satisfying, to find that nocturnal habits are the rule. Some gliders come out to bask in the sun and they can be seen airborne when disturbed, but they are generally active only at night. There is an exception, as always. This is the flying lizards. Whether they are the exception which proves the rule will not be known until the factors allowing their diurnal behaviour are understood.

Although their fauna is so rich, finding direct comparison between the day and night shift in tropical forests is not easy. There are the general comparisons between hawks and owls, fruit-eating birds and fruit-eating bats and so on, and one or two specific comparisons, such as the similarity in habits between the diurnal lesser tree shrew and the nocturnal pen-tailed tree shrew. There is, however, one group of animals which has diurnal and nocturnal members and has been studied intensively. This is our group, the primates. Excepting man, the order of primates is composed of two major subdivisions: the monkeys and apes, or anthropoids (literally 'man-like') and the prosimians (literally 'before the monkeys'). They are called, respectively, the higher and lower primates, and the lower primates are overwhelmingly nocturnal.

The prosimians comprise 33 species. In Africa there are seven species of bushbabies or galagos, the potto and the angwantibo; the orient has five lorises

The African bushbaby's nocturnal lifestyle keeps it out of the way of its monkey relatives.

and tarsiers but the real centre for prosimians is the island of Madagascar which has 21 species including many lemurs, the indris, two sifakas and the aye-aye. The number of species on Madagascar is significant. The prosimians were once abundant and widespread over the world but they have been largely replaced by the more modern monkeys and apes, except on Madagascar where the prosimians, lacking competition, have continued to thrive.

When Madagascar separated from Africa it was an ark carrying many types of animals to safety. In the same way that Australia became a refuge for marsupials while the placental mammals overran the rest of the world, so the prosimians were safe on Madagascar where there are few predators and little competition from rapidly evolving anthropoids and rodents. They have been free to evolve slowly along their own lines but elsewhere they have their backs to the wall. They have disappeared from North America and Europe and they have survived in Africa and Asia by adopting specialised ways of life to avoid their younger and more vigorous anthropoid cousins.

The prosimians have retained the ancestral mammalian nocturnal behaviour while the simians have become diurnal. It has been suggested that the original prosimians were nocturnal to avoid clashing with birds over fruit and insects, and that the anthropoids became diurnal after they had adopted new techniques for finding food. Their dextrous hands and quick minds enabled them to probe and peer in unexplored places. It is interesting that the only nocturnal monkey, the

120

The potto creeps slowly through the trees in search of insects and fruit.

night monkey or douroucouli, lives in South America where there are no prosimians and the only diurnal prosimians live in Madagascar where there are no anthropoids. They too have managed to exploit food which the beaks of birds and paws of other mammals cannot manage.

Some of the prosimians can be seen in zoos, where there are houses in which the day and night cycle has been reversed so that we can see nocturnal animals in their active period. They are rather charming creatures with round staring eyes and snubby snouts. Some are spectacular leapers with impressive acrobatic skills; others move dextrously along branches, with their feet clamping deliberately at every move.

The bushbabies of Africa range from the rain-forests of West Africa, where three of the five species live, to the open, dry bush of the East and South. They spend the day in a hollow tree or a nest of leaves, where the baby is left while the mother is away feeding, and they come out at night to feed alone or in small groups. Their diet is made up mainly of insects, fruits and the gums which are exuded from certain trees, but each species has its own preferences. The common bushbaby, which is found over most of Africa south of the Sahara and is often kept as a pet, prefers insects, but also takes plenty of fruit, and the thick-tailed bushbaby hunts larger animals along with eating more fruit. The dwarf bushbaby concentrates on insects and the needle-clawed bushbaby specialises on gums. It uses its pointed nails to cling head down on smooth tree

The trailing hindleg of this slender loris shows the pincer grip which aids its progress along slender stems.

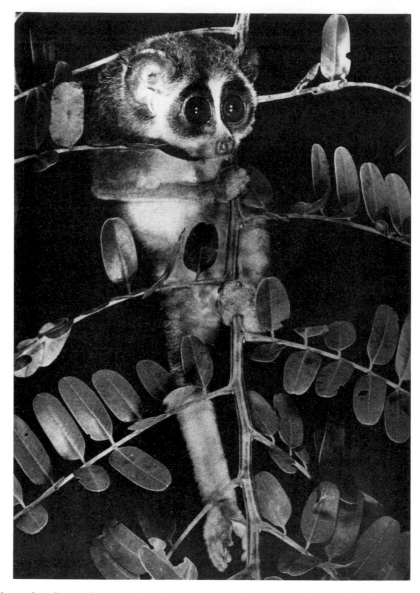

trunks and its lower front teeth—the incisors and canines—form a flat chisel to scrape up the gum.

Bushbabies are good jumpers. They take off and land in a vertical position and cross gaps of 5 metres. In this they differ from their closest relatives, the angwantibo and potto of Africa and the lorises of Asia, all of which creep slowly and deliberately along branches. Their index fingers and second toes are shortened to stubs giving a pincer-like grip. The angwantibo lives high up in trees, where it is rarely seen, and it is said to spend much of its time hanging upside-down. It once had the name of golden potto, and the true potto was called Bosman's potto after its European discoverer. Bosman could not have been

122

flattered to have the potto named after him because his original description told of its 'lazy, sluggish Nature; a whole day being enough for it to advance ten Steps forward' and that "tis impossible to look on him without Horror . . . . I don't believe anything besides so very disagreeable is to be found on the whole Earth'. Like the bushbabies, the angwantibo and potto have varied diets, but the former prefers insects and the latter fruits. Of the two lorises, the slender loris, named after its long, spindly legs, lives in southern India and Sri Lanka and hunts geckos and roosting birds as well as eating fruit. The slow loris lives in South-east Asia, from Assam to the Philippines and eats insects and fruit. It is peculiar in that it regularly eats insects which are distasteful to other animals.

The tarsiers, of which three species live in the islands of the East Indies, are among the most fascinating primates. They are tiny animals, about 100 grams in weight and about 35 centimetres long, of which over half is tail. Their erect ears, snub noses with sideways directed nostrils and huge round eyes make them oddly attractive, yet macabre. The eyes are so large that they are relatively immobile, but, like owls, tarsiers can turn their heads to look over their shoulders. The thumbs are not opposable, but the fingers are long and spidery and give a firm grip when the tarsier lands after a leap. It is propelled through the air by a frog-like thrust of its hindlegs which are lengthened by the elongated tarsus bone, hence the animal's name.

The lemurs and their relatives living on Madagascar have evolved in their isolation into a variety of forms and habits. They vary in size from the tiny mouse lemur weighing 60 grams and measuring only 13 centimetres overall, to the cat-sized ringtail and brown lemurs. These two are unusual in being diurnal and living sociably in groups. There are some nicely named lemurs: the gentle lemur, the variegated lemur, and the sportive lemur which earns its name from its habit of boxing its opponent with one hand. Some of these lemurs run and clamber but the indri and the sifaka are leapers, like the bushbabies and tarsiers. The indri is the largest lemur and the sifaka is the most impressive, with a coat of white, orange and maroon. The odd aye-aye, now reduced to a pitiful few, a course which other Madagascan prosimians may soon follow, is a weird-looking animal with a shaggy coat of black hair, a bushy tail and erect ears which lend it the air of a misshapen squirrel. Its main peculiarity is a long middle finger on each hand which it uses for getting its food. The aye-aye is the woodpecker of the prosimian world; although it eats mainly fruit, its speciality is insect grubs living deep in timber. It seeks them out by smell and by listening for their movements and digs a hole in the wood with its sharp incisor teeth. Then it squashes the hapless grub with its long middle finger and extracts the body juices by wiping them up with the finger and licking it. Where coconut plantations have been established, aye-ayes learnt to get the soft pulp of the coconut in the same way.

The prosimians have always been a rather mysterious group, living high up among dense vegetation and coming out at night. Little was known of their habits and their mysterious ways attracted plenty of folklore. The natives of Madagascar believed that a mere touch by an aye-aye caused death and that aye-ayes were the reincarnation of their ancestors. This situation has now changed radically over the last two decades as zoologists have undertaken intensive studies of primates living in the wild. For many zoologists, the main interest is the comparison of feeding habits and social life of different species and

The staring eyes of the tarsier are so large that they can hardly turn in their sockets.

124

how these facets of an animal's life are related to its environment, but prosimians are of especial interest here because they are one group of mammals which have been studied as nocturnal animals.

In the evolution of the primates, there has been an overall replacement of smell by eyesight as the main sense. The prosimians have advanced along that road to the extent that smell and vision are of equal importance, along with hearing. All the three senses are used by prosimians for finding food, tracing their way around their feeding grounds and communicating with each other. The potto can detect an insect hidden 1 metre away and smell offers a means of identifying prey at night in lieu of the colour vision which is used by diurnal species, although the eyes are still used for guiding the final attack. Secretions from glands on the body and urine are used for marking pathways but even nocturnal species seem to memorise visual features. The same scents are used for communicating identity and social standing to members of the same social group or to strangers, but facial expressions communicate moods, especially in the diurnal lemurs.

It is impossible for us to visualise the world of scent which prosimians construct around their environment but we can appreciate their calls which denote anxiety, alarm and reassurance, and we can begin to understand the visual world as seen by prosimians at night by considering the physical properties of their environment. The eyes of prosimians are of the dual-purpose type so they can see well by day, but all have got the tapetum which is typical of nocturnal animals. The exception is the tarsiers, which compensate with the large size of their eyes making them efficient at gathering light.

Due to the effect of the foliage, the light in a forest at night is mostly at the red end of the spectrum. Prosimians, and other nocturnal animals, are relatively insensitive to red light. This is why it is used for observing animals at night. However, light coming from the night sky is bluish and matches the peak sensitivity of prosimian eyes. It acts as a spotlight throwing objects into contrast against the dark red backdrop of the forest. This phenomenon offers an explanation for the observation that lemurs living in dense, humid forest have reddish fur compared with the greyish fur of inhabitants of open, drier woodland. They will merge with the background, as nocturnal predators have a similar shift in sensitivity to the blue end of the spectrum.

Deserts

Deserts stretch in a broad belt around the world and, through a complete lack of foresight on the part of man, which allows overgrazing and wholesale felling of woodland, they are spreading at a disturbing rate. The proportion of the world's land surface which is desert is alarmingly large. The Great Palaearctic Desert stretches from the shores of the Atlantic Ocean on the west to the heart of China in the east. It encompasses the Sahara of North Africa, the Arabian Desert, and large tracts of Central Asia, including the Gobi and Thar Deserts. The second largest desert is in central Australia and other deserts include the Kalahari and Namib of South-west Africa, Patagonia, the Atacama Desert of Chile and Peru and the deserts of south-west United States and north-west Mexico. Not all deserts are hot all the time; some, like the Gobi, have bitterly cold winters. The popular image of a desert is of a barren stretch of sand, perhaps thrown into wave after wave of immense shifting dunes and broken by angular, rocky hills. Such barren wastes, sometimes made up of gravel and pebbles or bare, wind-eroded bedrock instead of sand, are found in the heart of deserts. Around them, there are slightly less arid places supporting sparse growths of hardy, drought-resistent thorny trees and bushes, and in America, cacti.

By definition, a desert receives very little rain and that erratically. Thus, it is not unusual for a year or more to pass without rain in some places. When rain eventually does come, it falls as a deluge which even bone-dry soils cannot absorb and the countryside becomes inundated. Wildlife is immediately activated. Frogs and grasshoppers start to call in deafening chorus and long-dormant seeds germinate. Except in the most barren places, a shower of rain brings a rapid growth of ephemeral plant life which flowers, sets seed and dies before the ground dries up again. During the short period of abundance native animals are joined by nomadic animals, especially birds and locusts, which arrive to share the bounty. Yet, even in the driest places, there is some animal life which ekes a living on plant material blown in by the wind. Around the margins of the deserts, there are semi-arid zones with a more regular rainfall and easier conditions for life, but still inhabited by animals which need to survive hot weather and drought.

As dwelling places for animals, deserts offer a marked contrast to the tropical forests of the previous chapter. The environment inside the forests is warm and moist and provides good conditions for delicate animals. It is also equable, with little variation in physical conditions over the 24 hours. Deserts lack the stabilising effect of the vegetation; they are hostile overall and subject to violent fluctuations in temperature and humidity. The principal interest for the biologist working in a tropical forest is the manner by which a dazzling array of species have adopted narrow specialisations of habit to allow them to flourish together without undue competition. The fascination of deserts, on the other hand, is the way that a few species have evolved to cope with a hostile environment. Compared with the riot of adaptations seen in the forests, unrelated desert species from different parts of the world have evolved similar adaptations for avoiding heat and conserving water. One such adaptation is to confine the period of activity to the hours of darkness and in no other environment is there such a clear advantage for a night life.

The dry atmosphere of deserts, with cloudless skies and very low atmospheric humidity, results in the sun's rays striking the ground with great intensity. Recorded ground temperatures have exceeded 80°C but the clear air also allows the heat which the ground has absorbed during the day to radiate rapidly back into space at night, so the temperature at ground level can fluctuate by as much as 50°C between midday and nightfall. However, within a few centimetres either side of the surface, the fluctuation is damped down and animals can escape the heat by burrowing under the surface, by flying above it or even by standing on long, stilt-like legs.

The graphs show the varying air and soil temperatures throughout the day and night.

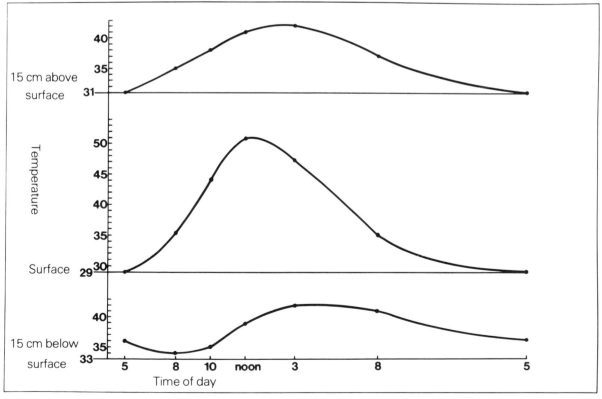

Sand dunes at dusk. The few
animals that live here will come
out at night when it is cooler.

The effect of high temperatures on animal life is exacerbated by the low
humidity, which increases the evaporation of body fluids. At night, the fall in air
temperature leads to a dramatic rise in relative humidity, from approaching zero
to 40–50 per cent and dew may be deposited on the ground. As soon as the sun
rises, the dew evaporates but its brief appearance is a great benefit to animals. By
feeding at night, grazing animals can get much needed water. Where the desert
is near the sea, onshore winds bring in moist air and mists form. In the Namib
Desert of South-west Africa, mist and fog are sufficiently frequent to bring life to
what would otherwise be completely dry dunes.

To survive in the desert, an animal has to avoid overheating and drying up.
The two problems are related because heating increases the rate of evaporation
and a common physiological mechanism for keeping cool is to allow the body
water to evaporate. Human beings keep cool by sweating and kangaroos lick their
fur so that evaporation of the saliva can cool them. Evaporation and the use of
body fluids to keep cool is allowable only if the animal has access to a water
supply and can afford to lose some of its body water. The problem of water loss is
acute for animals feeding on dry plants but less so for carnivores and insectivores
which take the fluids of their prey, and for animals subsisting on succulent plants.

The solution for many desert animals is to escape into the night. They spend
the day in burrows, where the atmosphere is cool and moist, or under stones and
in the shade of plants, and become active after nightfall. Small animals, whose

128

Scorpions live in many places but some are well-suited for life in deserts. They avoid the heat of the day by burrowing.

relatives in temperate regions and tropical forests may be out by day as well as night, are strictly nocturnal in the deserts. Desert woodlice, millipedes and insects have a more impermeable cuticle than those from moister habitats, but they still take advantage of the favourable conditions at night. Only a few of the desert representatives of these animals manage to be active during the day. The most conspicuous are large black darkling beetles which crawl over the ground on long spindly legs that hold them clear of the scorching ground, but even the few diurnal insects tend to become crepuscular in the hottest weather.

The scorpion is one of the most characteristic desert animals. Some live in forests but as the group they are well adapted for life in deserts. The body has a thick impermeable cuticle and, like their spider relatives, they subsist on a liquid diet. Scorpions spend the day under stones or in crevices. Some dig their own burrows or shelter in those of rodents. While studying sand rats in southern Tunisia, I always wore gloves while excavating their extensive burrows after I had discovered scorpions lurking inside. How the sand rats avoided fatal encounters, I do not know. The scorpions came out at night and we also learned to examine our boots and clothes in the morning in case they sheltered a scorpion. The Tunisians were more scared of scorpions than we were and the local forester would not bring his family from the town to join him. This is a well-founded fear. Scorpions are responsible for numerous deaths and a sting from some species can kill a man in a few hours. The usual prey is insects and spiders which are tracked down mainly by touch and seized in the massive pincers.

129

Thick-skinned darkling beetles are one of the few kinds of animals which can survive in the desert sun.

Along with the scorpions and spiders, there is a third group of eight-legged arachnids, whose appearance is just as unprepossessing. The camel-spiders, false-spiders or wind-scorpions as they are variously called, look rather like spiders, but they have huge jaws and a thin covering of long bristles. They are nocturnal, their scientific name Solifugae meaning 'hiding from the sun', and they are extremely fleet of foot. A camel-spider at speed looks like a ball of fluff blown by the wind. The largest are 15 centimetres across the limbs and repulsive to the eye although harmless to humans. Their prey is any animal which can be overcome and ranges from insects to lizards, small mammals and birds. The victim is grabbed and crushed to a pulp by being run through the camel-spider's jaws. Even scorpions are not immune from attack, and a duel between a camel-spider and a scorpion becomes a contest of fast reflexes. The scorpion attempts strikes with its sting, and the camel-spider dances like a boxer to evade the sting and leaps in to bite and hold tenaciously to the scorpion's tail until the sting is severed.

The reptiles as a group are well adapted for survival in dry conditions. Many prey on insects and other small animals or, like the tortoises, feed on succulent plants and are independent of free water. Most, like the tortoises and the majority of lizards, are diurnal. They avoid the greatest heat of midday but otherwise they rely on the sun's warmth to maintain a high body temperature and they bury themselves at night in warm sand to conserve heat. The exceptions to this regime are found among the geckos and snakes. Lizards are mainly diurnal but the majority of desert snakes, which prefer lower temperatures than lizards, are nocturnal. As in the forests, most geckos are abroad at night, and the deserts of south-west United States and Mexico are the home of two more mainly nocturnal species which are famous as the only venomous lizards. The Gila monster and the bearded lizard are sluggish and eat slow-moving insects, or rob bird's nests, which they track down by smell, using the forked tongue to transfer scents to Jacobson's organ. Their venom is quite powerful and human fatalities have been recorded but the fangs are small and a lethal quantity of venom is injected only if the lizard can hang onto its victim and chew its flesh.

The pit-vipers are notably tolerant of low temperatures and their heat-sensitive organs give them the means of hunting in the dark. They spend the day in rodent burrows or sheltering in the shade of rocks and bushes, and they are dangerous only when disturbed. Luckily, they take some time to arouse. The rattlesnakes are pit-vipers which are found from Canada through the United States and Mexico, and into South America. Some live in well-timbered country but they are best known as desert animals from their appearance in 'cowboy' films when they are brought in to add a further hazard confronting 'our hero'. The snake conveys its annoyance at being disturbed by shaking the rattle at the tip of its tail. This is made up of a series of interlocking scales.

Several rattlesnakes live in the deserts and semi-desert country of south-western United States and Mexico, where they can be such a threat to human and animal life that organised hunts are carried out. The sidewinder is a specialised desert-dweller, named after the method of progression which it shares with other desert snakes, such as the horned viper and puff adder of Africa. Sidewinders are most active in the early part of the night when the air is still quite warm. They live in rather barren areas where there is loose, windblown sand and their 'sidewinding' method of locomotion gives them an advantage over other

The horned viper of African deserts shows the slit pupil which is typical of nocturnal animals when seen during the day.

131

The unprepossessing Gila monster is one of only two lizards which are venomous.

Opposite: An eight-legged relative of the spiders, the camel-spider's scientific name of *Solifugae* means 'hiding from the sun'.

snakes. The tail of a sidewinder consists of a series of parallel lines, each with a hook at one end made by the tip of the tail. Instead of the usual eel-like wriggling in which a series of waves pass down the body and push against the ground to drive the snake forward, the sidewinder is propelled by a rolling action rather like a caterpillar tractor. The snake folds its body into a series of tight curves and raises itself so that only two parts are touching the ground at one time. As the snake moves, the points of contact move down the body so that it is pushed forward. When one point reaches the tip of the tail, a new point is started at the head end.

The deserts of the world are inhabited by a remarkable number of small rodents, which look alike but are not closely related. They have adapted to the desert environment by evolving along the same lines. In North America, there are the kangaroo rats, desert pocket mice and spiny pocket mice. The great sweep of deserts in the Old World have the greatest variety, with over 50 species of gerbils and jirds, steppe lemmings, desert dormice and jerboas. Australia has its own kangaroo rats, of a family unconnected with the American kangaroo rats, and it has the jerboa marsupial, crest-tailed marsupial mice and pouched mice, which are desert-living marsupials similar to the rodents.

These small mammals have a number of traits in common, which have evolved in parallel as a result of their desert lifestyles. Their sandy coats render them inconspicuous; they tend to have hairy feet to grip loose sand and strong

133

The sidewinder, or horned rattlesnake, is one of several desert snakes which use the 'sidewinding' method of locomotion to travel over loose sand.

claws for burrowing; and they have sensitive hearing. Many progress by hopping as their names suggest and they have long tails, often with a tuft of hairs at the tip. They can survive extreme drought and a few need never drink. Some eat insects; other feed on plants, often dry seeds. Almost all are nocturnal and spend the day in a burrow.

The gerbils and jirds of the Old World look rather like rats but are, in fact, related to the voles and hamsters. The many species go by different names, such as fat sand rats, fat-tailed mice and naked-soled gerbils. They live in deserts and semi-desert scrub lands, and several have become well-known as pets and laboratory animals. Some gerbils, like the fat sand rats I found in Tunisia, make extensive burrow systems with several entrances and side-chambers in which to store food and build nests of shredded leaves. The food stores are stocked during times of plenty when rain has freshened the ground and there are leaves and seeds in abundance. The great gerbil stores as much as 50 kilograms of food, and gerbils wreak greater havoc in the poor agricultural land of the scrub regions than their numbers would suggest because they carry away to store more than they can eat at the time. A few gerbils are carnivorous and eat locusts and other insects. Other desert insect eaters include the several desert hedgehogs in North Africa and Asia, the grasshopper mice of American deserts and the mulgara or crest-tailed marsupial mouse of the Central Desert of Australia. They secrete a concentrated urine, so they can rely on their food providing sufficient water for their fluid-conserving nocturnal regime.

134

By spending the day deep in a burrow and conserving their body fluids, gerbils can survive long periods without drinking.

Gerbils have long back legs and, while they walk or scamper on all fours at leisure, they travel by bounding when in a hurry. The Indian gerbil is also called the antelope rat because it covers the ground in bounds of 5 metres. There has been a general tendency for desert rodents to move by bounding on all fours or hopping like kangaroos. It is thought that this is an economical way of moving about in search of scattered supplies of food. The rodents can also elude predators by jinking violently.

The development of a hopping gait reaches its peak in the jerboas, which look like tiny kangaroos and are the 'desert rats' of World War II. At night, they can be picked out in headlights by their reflective eyes which bounce erratically in long arcs out of the path of the oncoming vehicles. The jerboas' head and body lengths range from 4 to 15 centimetres, with disproportionately long hindlegs of 3–8 centimetres and tails which are longer than the rest of the animal. The size of the legs is due mainly to the length of the feet in which the toe bones are fused, like the cannon bone of cattle, for extra strength. The front legs are small and, like kangaroos, jerboas move on all fours only while feeding. Otherwise they 'trot' in short jumps of 10–12 centimetres and 'sprint' in long bounds of 3 metres.

A good sense of hearing is a feature of desert animals and may be related to their nocturnal habits. Some jerboas have large ears and look like miniature hares. The long-eared jerboa has ears half as long as its head and body which can be folded up as it enters its burrow. Others have small external ears like voles but

their auditory bullae are enlarged. The bullae are hollow, dome-shaped
chambers at the base of the skull which enclose the middle-ear mechanism. The
dwarf jerboas have huge bullae which swell the size of the skull so that the head
is as big as the body. Enlarged bullae are also found in some of the gerbils and
the American kangaroo rats and mice, as well as in desert-dwelling antelopes,
foxes and bats.

In the American kangaroo rats, the bone of the bullae is paper thin and they
make up a volume larger than the rest of the cranium. The advantage of the large
bullae is that they do not restrict movement of the eardrum and the three tiny
bones of the middle ear which transmit sounds from the exterior to the sensory
tissue of the inner ear. The kangaroo rat's ear is especially sensitive to sounds of
frequencies between 1 and 3 kHz, a range to which most small mammals are
rather insensitive. In an experiment, the bullae of some kangaroo rats were partly
blocked with wax, in such a way as not to interfere with the ear mechanism, and
their sensitivity was found to be very much reduced. The value of this sensitivity
was then demonstrated by keeping kangaroo rats in a lightproof room with either
a barn owl or a rattlesnake. In these conditions the owl hunts by hearing and the
snake uses its heat-sensitive pits, but each time the one pounced or the other
struck, the kangaroo rat leaped straight up and landed out of danger. Next, the
bullae were blocked to reduce the ears' sensitivity and the unfortunate rodents
were rendered helpless with no warning of the impending attack. To set the cap

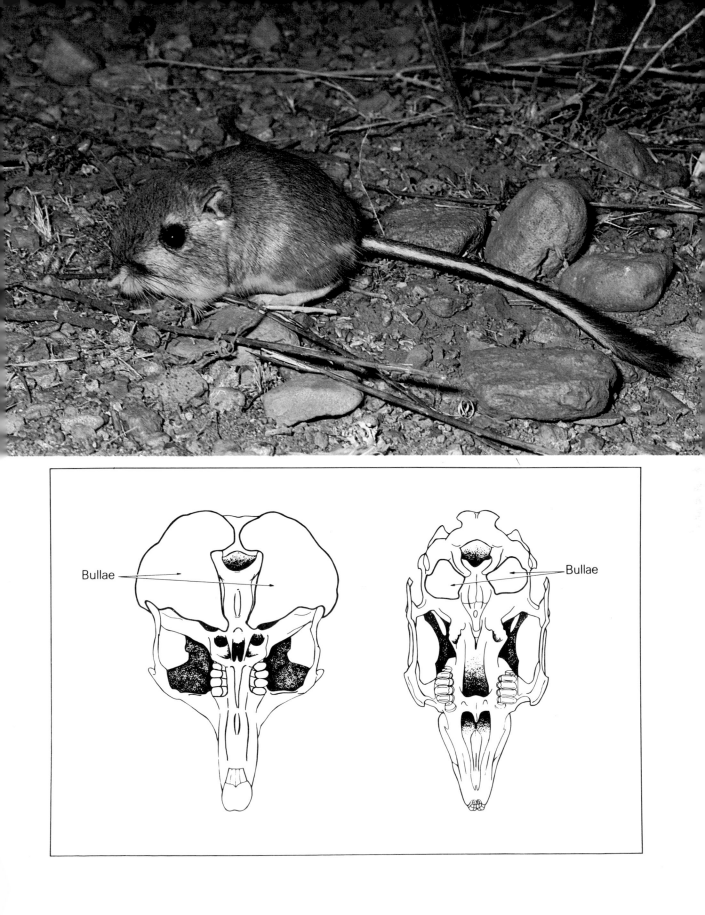

Bullae

Bullae

on the story, a tape recorder picked up sounds, far too faint for the human ear, made as the predators struck. The owl's were of 1.2 kHz and the snake's were of 2 kHz, both within the ultrasensitive range of the kangaroo rat's hearing.

If the external ears are a good indication, the rodents are not alone among desert mammals in having a well-developed sense of hearing. Two of their predators have prominent ears. The fennec, or fennec fox, is a delightful little animal, about the size of a Jack Russell terrier but with ears 15 centimetres long. Fennecs live in the Sahara and Arabian Deserts, spending the day in extensive burrows and emerging at night to feed on insects, spiders, scorpions and more rarely rodents, snakes and lizards. The bat-eared fox of the arid regions of East Africa and southern Africa looks like a lightly built red fox but with largers ears. It hunts mainly insects, especially termites and its rather weak teeth recall the state of the teeth in other termite and ant-eating mammals, such as the aardwolf. Gerbils are sometimes hunted by the bat-eared fox which is as skilful at jinking and doubling back on its tracks as the gerbils.

The survival of kangaroo rats and jerboas in all but the most arid and barren deserts is due to their ability to subsist with the minimum of water. Kangaroo rats and some jerboas live on dry seeds yet they survive indefinitely without drinking. They obtain a small amount of moisture from their food, and water is a by-product of the oxidation of sugars and fats to release energy in the body. This is sufficient to maintain the body fluids if water loss is cut to a minimum. Water is lost from the body in sweat and excretory products and in exhaled breath but

The jerboa can survive indefinitely on the dry seeds without drinking through several physiological adaptations for conserving water.

rodents do not sweat and desert rodents produce highly concentrated urine and dry faeces. The nose is designed to reduce the amount of water vapour lost in exhaled air. When cool air is inhaled, it cools the lining of the nasal passages. At exhalation, moist air at body temperature passes back and is cooled in the nasal passages sufficiently for some of the moisture to condense. This phenomenon only works when the atmosphere is relatively cool, a situation which occurs by day in the rodents' burrows and at night in the open.

This phenomenon alone will explain the desert rodents' regime of holing up by day and feeding at night, but, even with a plentiful supply of drinking water, they cannot survive for long in the sun. As a consequence of an elementary law of geometry a small animal has a larger surface area proportional to its volume than does a big animal. A small animal will therefore heat up (or cool down) more rapidly than a bulky animal, and the latter can afford to keep cool by evaporation because it has a greater reserve of body water proportional to its surface area. If desert rodents are forced out of their burrows during the day they soon overheat, but sweat glands would be no help if they possessed them because they would rapidly die of desiccation. In an emergency, however, jerboas can keep cool by salivating over their fur.

To contradict the conclusion that it is vital for desert rodents to be nocturnal, a few species can be seen by day but these live on the fringes of deserts where conditions are not too extreme and they have developed protection from the sun. The midday jird has extremely thick fur and thick skin and the great gerbil has heavily pigmented skin to shield it from harmful radiation.

Animals up to the size of foxes can escape the heat of the day by burrowing but larger animals have little chance of escaping the full heat of the sun. Nevertheless, their large bulk offers some protection because they will take longer to warm up and a thick coat of hair or fur keeps the sun off the animal's skin. The camel and oryx allow their body temperatures to rise to reduce the need to sweat

The bat-eared fox is one of the largest desert animals which can live in a burrow by day and come out in the cool of the night to hunt gerbils.

and a camel does not sweat until its temperature has risen to 40.7°C. When possible, desert animals make the most of the shade offered by trees and boulders. While resting, they will generate less heat in their muscles and reduce the heat load on the body. Furthermore grazing at night has the added advantage that the herbage will be dampened with dew. Oryx, for instance, can live in areas of dry scrub without the need to drink.

The natural history of Australia and its strange collection of marsupial mammals must not be overlooked here. In the middle of this huge continental island there is an extensive area of desert and semi-desert, which is inhabited by a number of animals. Most marsupials are nocturnal, the wombat being an exception, so their daily regime suits them for a desert life. As well as the small mulgara, the dunnart or pouched mouse, and the jerboa marsupial, some of the largest kangaroos are to be found in dry places. The red kangaroo and the wallaroo or euro rest during the day, the wallaroo sometimes sheltering in caves, and they graze or browse by night.

The evidence that desert animals are nocturnal to escape from the sun's heat is clear-cut, but John Cloudsley-Thompson, formerly professor at the University of Khartoum and an expert on desert life, has stressed the role of nocturnal activity among desert animals as being also a means of escape from predators.

As in other places, the birds of prey and various insect-eating birds, such as the shrikes, are important diurnal predators, along with many lizards, with the exception of the geckos. The distribution of birds is limited by the availability of water and they feed during the cooler parts of the day. Small birds shelter from the hottest sun in trees or under boulders, while larger birds of prey soar high overhead where the air is cooler. The lizards—agamids and lacertids in the Old World and iguanas in the New—make use of the sun to maintain a high body temperature at the beginning and end of the day but avoid the worst heat. Some feed on plants; others on small animals, and the 2-metre-long desert monitor of Africa chases and devours small mammals or anything else it can overwhelm.

Balanced against the diurnal predators, there are the nocturnal snakes, geckos, and insect-hunting rodents, and most of the carnivores hunt at night. The desert and semi-desert regions are inhabited by a variety of carnivores, from the specialists like the fennec, bat-eared fox, the similarly long-eared American kit fox and the African sand cat to wide-ranging, adaptable hyaenas, bobcats, caracals, red foxes and wolves. Carnivores, as a rule, have flexible habits and hunt when they have the best chance of success, but with so many prey animals confining their activity to the night, it is not surprising to find that the hunters are also nocturnal and they, too, will no doubt be avoiding the heat.

Cloudsley-Thompson points out that scorpions and camel-spiders can survive the most intense desert heat without difficulty so hiding by day must be a matter of defence. Despite formidable stings, scorpions fall prey to a number of predators, including baboons, which nip off the stings with their fingers. However, as baboons search under stones and delve into crannies in their search for food, scorpions will not be significantly safer when they are taking refuge. Without having much experience of the desert, and from the comfort of my study, I favour the third explanation that scorpions come out at night at least partly because this is the time their own prey is active. Nevertheless, this does not explain fully why scorpions are so rigidly nocturnal, from forests to deserts.

141

The Seas

In the heyday of the herring fishery, during the early years of this century, the seas around northern Europe, twinkled with dancing lights at night as hundreds of drifters shot their nets to intercept the gathering shoals of herrings. The fleets of drifters have disappeared, but they once set sail every evening from small fishing ports scattered along the coast and returned home next morning with holds crammed full. The shoals of herrings appeared off different parts of the coast through the year and the fishing industry moved with them. For instance in Britain, June and July saw the fleets in action around Scotland, from Mallaig in the west, through Lerwick in Shetland to Peterhead on the east, but by October the main harvest had come to the East Anglian ports of Great Yarmouth and Lowestoft.

Wherever they worked, the fishing fleets operated by night. Each boat streamed an immense net, 4 kilometres long and hanging from a row of buoys like a curtain 15 metres deep. The herrings were caught near the surface of the sea, and the drifters worked unsocial hours because the floating nets would catch nothing by day. The herring spend the day at some depth, come up in the evening and retreat next morning and the reason for this commuter traffic is the daily vertical migration of the planktonic animals on which the herring prey.

As soon as oceanographers began to study planktonic animals methodically, they found that there is a regular up-and-down movement through the 24 hours with their catches of plankton near the surface being much larger at night. This vertical migration of plankton is a general phenomenon in salt water and fresh water, and involves a wide range of very diverse planktonic animals: crustaceans, arrow-worms, the medusa larvae of sea anemones and jellyfish, sea butterflies, the young stages of fishes, comb jellies, salps and many others, even single-celled protozoans.

The facts of vertical migration were studied in detail by using a small copepod crustacean called *Calanus*. This is about the size and shape of a grain of rice, and has a short tail. It hangs vertically in the water with two long, bristly antennae outspread to slow its rate of sinking and, at intervals, it beats its second pair of smaller antennae to maintain a constant depth. At midday, the bulk of *Calanus* are hovering between 15 and 30 metres below the surface. During the afternoon they start to ascend and, by nightfall, most of them have gathered

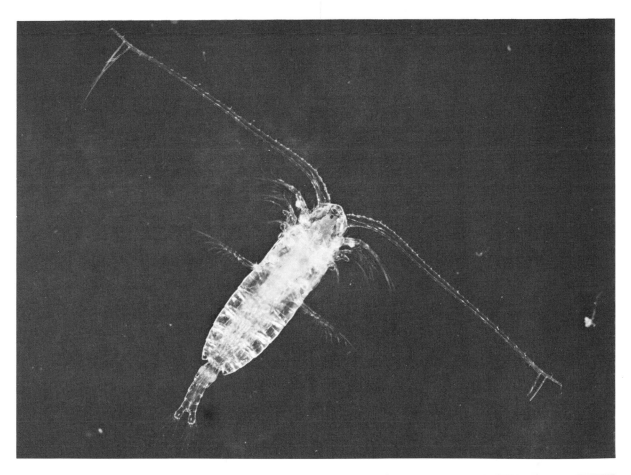

Above: *Calanus* is a small crustacean which lives in swarms and is the food of many larger marine animals.

Right: A variety of phytoplankton, the microscopic floating plants which feed the smallest marine animals.

143

within the top 5 metres of the sea. They then remain near the surface, with perhaps a short descent and return during the night, and finally return to deep water at daybreak.

The general pattern of movement by *Calanus* and other planktonic animals was demonstrated by trailing nets at different depths behind a ship, but to study the extent and rate of movement, individuals had to be caught and brought into the laboratory. There they were observed in a 'plankton wheel'. This is the size and shape of a cartwheel, its rim being a transparent perspex tube filled with sea water. An animal, in this instance a *Calanus*, is placed in the tube and the wheel rotated until the animal is at convenient eye-height for the observer. As the animal swims up or down, the wheel is rotated, keeping the animal stationary relative to the observer, and the amount of rotation is recorded. A series of trapdoors within the tube ensure that the water rotates with the rest of the wheel, so the animal is, in effect, in an endless column in water. A minute animal like *Calanus* was found to swim upwards at a rate of 15 metres per hour and downwards at 100 metres per hour. Larger planktonic crustaceans achieved much faster rates.

The almost universal occurrence of vertical migration among the assembly of quite unrelated animals in the plankton, its occurrence in freshwater lakes and ponds as well as in the sea, and the amount of energy which must be consumed in swimming up and down, sometimes over tens or hundreds of metres, lead to the inevitable conclusion that vertical migration must be of fundamental importance to the lives of these aquatic animals. The widespread nature of this phenomenon was realised over 100 years ago, but the reason for vertical migration remained a mystery. Indeed, Sir Alister Hardy's *The Open Sea*, the classic account of the natural history of marine plankton, published in 1956, has a chapter entitled 'The puzzle of vertical migration'.

The attraction of the upper layers for planktonic animals is the mass of phytoplankton—the minute floating plants—on which they depend for food. Phytoplankton must, of necessity, live near the surface where there is sufficient sunlight for the essential process of photosynthesis, so zooplankton have to come to the surface to graze. The puzzle was why the mass of animals actively swam away from their food during the day.

Several answers to this conundrum have been suggested and the consensus is that the principal reason for vertical migration is that planktonic animals feed in the phytoplankton-rich surface waters under the cover of darkness and then actively swim down to spend the daytime hours in the gloom of deeper waters to avoid predation. If they remained at the surface, they would be more easily found by plankton-eating fish and squid.

The reaction of zooplankton to light is further demonstrated by the habit of *Calanus* of coming nearer the surface in foggy weather or when the amount of light penetrating the sea is reduced by the effect of wind or rain on the surface. The importance of predation on their habits is emphasised by the way young stages of crustaceans, too small to be eaten by fishes, remain at the surface. The immunity conferred by darkness is, as usual, not complete, otherwise the nightly surfacing of herring would be pointless. The planktonic animals are still

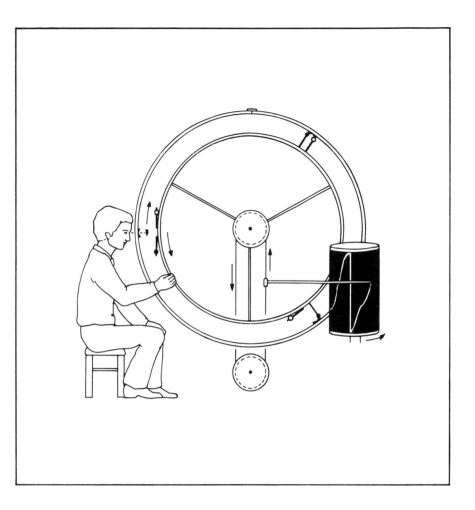

The 'plankton wheel', a device for recording the vertical migration of small animals. As the animal moves up or down, the wheel is revolved to keep it at eye-height and the distance the wheel travels is recorded to give the distance the animal swims. The trapdoors ensure that the water inside the wheel moves with it, but does not hinder the movement of the animal. (*After* Hardy, *The Open Sea*, Collins).

consumed in large numbers but each individual has a better chance of survival than it would have by remaining at the surface all the time, where they would be eaten not only by herring but also by many other marine creatures.

Some deep-sea fishes, notably the deep-sea smelts and lantern fishes, as well as squid, come up from considerable depths to feed at night. These, in turn, attract the attention of diving seabirds and dolphins. Observations of dolphins fitted with radio transmitters have shown them making shallow dives of no more than a few metres depth during the day and then descending to around 40 metres after nightfall. At the same time, echo sounder traces show that the deep-sea fishes are moving up to this depth themselves to feed on some shoaling organism.

If the behaviour of planktonic animals is geared to reducing the chances of being seen by predators, a new puzzle is presented by at least some of them. Many kinds of marine organisms carry their own source of illumination, including in the Southern Ocean, the crustacean popularly called krill. This lives in huge swarms and forms the main diet of whales and many seals, birds, fishes and squid. By day the shoals of krill stain the surface of the sea pink, while

A squid with a fish that it has caught. Squid follow the general movement of animals towards the surface at night.

Opposite top: Krill is the staple food for whales, seals, penguins and other animals in the Antarctic. Like many small marine animals, krill lights up in the dark.

Opposite bottom: Lantern fishes live several hundred metres down but swim to the surface at night. Their bodies bear patterns of light-producing organs which are used for recognition.

at night it illuminates the sea so brightly that the animals can be seen in their own blue-green light, each krill carrying a row of light-producing organs on each side of its body. So krill seems to be doing nothing to avoid predation, apparently surviving by sheer weight of numbers.

Each light-producing organ is like a miniature electric torch with a lens and reflector enclosing the source of light, which is generated by biochemical action. The scientific name for krill is *Euphausia superba*. *Euphausia* comes from the Greek words for 'true' and 'shining light' and a shoal of illuminated krill is superb indeed. Yet these lights, far from seeming to benefit the krill, seem to provide the krill's many predators with a sure target. Krill are certainly not distasteful like the fireflies. Not only are they eaten by many other animals; they are now being marketed for human consumption!

The ability to light up, or luminesce, is found in many marine animals from the protozoans to the fishes. One scientist has described the phenomenon as being so widespread that its absence in any one group of marine animals is more remarkable than its presence and that luminescence is probably as important in the oceans as coloration is on land. It is most spectacular in the tropics where the entire surface of the sea glows with luminescent bacteria or sparkles with pinpricks of light coming from tiny unicellular dinoflagellates floating at the surface. The dinoflagellates are goaded into lighting up by mechanical stimulation and a bucket of seawater will sparkle when stirred. A ship passing through these waters sets up a trail of light on each side as its bow wave spreads out, and

146

the tops of the waves are illuminated as far as the horizon. The most spectacular displays occur in the Indian Ocean when 'luminous balls' erupt from the depths and 'explode' into a disc of light which expands to about 100 metres in diameter before gradually fading. At other times waves or circles of luminescence travel rapidly across the sea or rotate like huge wheels.

Lesser displays can be seen in temperate seas. I have waded through the incoming tide and delighted in seeing rings of light scattering with each movement of my legs. On one occasion, late at night, the newly drained beach was peppered with glowing lights which came from the remains of small jellyfish, or some similar animal for only shiny globs of jelly remained. By dipping a finger in the jelly we could write our names in fiery letters on the breakwaters.

A score of reasons have been suggested for luminescence in marine animals, but despite a growing knowledge of the mechanism, its functions are still almost unknown. While the flashing of fireflies has been the subject of close study, observations at sea have been difficult and it has been rarely possible to do more than make conjectures about the role of luminescence in the lives of marine animals. This is yet another example of a subject where interpretation of function is still held up through lack of facts. Observations to date are comprehensive enough to show the scale of the phenomenon to be bewilderingly broad but insufficient to reduce it to order. The most likely functions can be summarised as: luring prey, defence, and social behaviour.

A well-known example of prey capture is the luminous lures of deep-sea angler fishes which they dangle over their mouths to entice prey within range. Deep-sea anglers may come to the surface to feed at night and one was seen using its light organs to illuminate its prey, which in this instance was krill! The fish was shining a beam of bluish light and by tilting its body upwards it illuminated a swarm of krill above it. It then swam up to snap at them. This sounds unnecessary if krill are already glowing with their own light.

One of the main dangers facing animals living near the surface is that they are silhouetted against the background brightness above them and so are very visible to predators beneath. Some predatory fish make a habit of attacking their prey from below and the prey's countermeasure is to reduce the contrast between its silhouette and the background. This is the function of the silvery scales of fishes which live near the surface over deep water. Each scale acts as a mirror so that the fish reflects, and therefore matches, its background from whatever angle it is approached. Light organs found on the undersides of some other fishes, as well as squid and crustaceans, may serve the same purpose: to balance their illumination with sunlight filtering down so that the animals effectively disappear.

Camouflage cannot be the answer for the illumination of swarms of crustaceans, which appear more as gigantic beacons calculated to attract predators. These sometimes show patterns of light which are unique to each species. Moreover, they may flash in synchrony. Here, the light probably serves to keep members of a species together and to control their rate of movement in vertical migration. This could be the function of luminescence in krill. Contrary to what seems logical at first sight, bunching together is an effective defence, not

148

Silversides look like Siamese twins when the camera catches their reflections as they feed at the surface by night.

only for animals armed with hoofs or horns, but for those which are at the mercy of predators. Forming a shoal, flock or herd may not reduce the number of animals eaten by a predator but it means that each individual has less chance of being killed and, in addition, predators may be confused by an embarrassment of prey.

The behaviour of fishes at night is not easy to study and the history of the Caribbean flashlight fish is an extreme example of the difficulties. It lives at depths by day and comes up at night to feed on shrimps and fish, which it traps in the beams of two powerful light organs situated under the eyes. The first specimen of a flashlight fish was found floating on the surface of the sea in 1907, and nothing further was seen of the species for another 70 years. It is now spotted regularly by divers, but only when they use extreme stealth. The flashlight fish is very wary of light; even the light of the moon stops it from emerging from the depths and divers can find it only if they descend without lights.

Other species of flashlight fishes live in the Indian Ocean. All are as strictly nocturnal and averse to light as the Caribbean species. They are equipped with powerful lights, which not only can be rotated to illuminate prey, but also probably attract shrimps which have the same in-built attraction to lights as moths. One species of flashlight fish found in the Red Sea uses its lights to confuse predators by suddenly disappearing. When alarmed, it shuts off its light, swims aways on another course, then lights up again. Flashlight fishes also use their lights to communicate with each other. They flash them on and off rapidly by 'blinking' with a shutter over the eye on meeting and, when a strange flashlight fish comes too close to a pair's territory, the female switches off, swims up to the intruder and lights up again. This is a signal for the intruder to retreat.

Opposite: The tentacles of Golden *Tubastrea* coral unfurl at night to catch passing animals. During the day, the coral is withdrawn into its skeleton.

Top: Fishes living around reefs are active in shifts. These cardinalfish are active by day. They hide in the reef at night while other species are on the move.

Bottom: The predatory moon wrasse hunts at night.

Luminescent fishes are unusual in shallow water but the midshipman is an exception. It is a slow-moving bottom-hugging species with a heavy head and large mouth set with sharp teeth, and it spends the day buried in mud. The underside of the body bears an array of several hundred light organs which produce light bright enough for reading a printed page. The light is used to scare predatory fishes such as groupers, and by both sexes in courtship.

Night diving has shown the patterns of activity among many fishes living in shallow water. Populations of fishes living in or around reefs of coral or rock show a distinct division into nocturnal and diurnal shifts. The nocturnal species typically have large eyes, as is to be expected, and are often coloured red. As seawater absorbs red light well, a red fish will be difficult to see in dim light. So we find that they spend the day lurking in crevices and caves or gathering in dense shoals near the bottom and spread out to feed at nightfall, when they are less easily seen.

The division into nocturnal and diurnal shifts seems to be partly a means of reducing competition by the two species involved sharing a source of food, as in the case of the cardinalfish and the damselfish which live on the same reefs in the Gulf of California. Both spend their active period in loose shoals hovering over outstanding rocks, but the former takes up position by day and the latter by night. There is an overlap at dawn and dusk when they intermingle. This is unusual because twilight is normally a time of quiet on the seabed, as one shift has already retired and the next has yet to appear. Both are avoiding the fish-eating groupers and jacks. These large carnivorous fishes hunt at any time of day or night when prey is available but dawn and dusk are particularly propitious because the dim light renders them inconspicuous against the bottom while still showing up the prey above them. The leopard grouper of the Californian reefs captures most of its food from shoals of herring which move to and from their offshore feeding grounds at this time, when they are more visible against the relatively bright surface. For the hunters, there is a fine balance between having

Shore crabs hide under seaweed and stones by day and move out at night when seabirds and other hunters will not find them so easily.

152

Hatchling turtles leave the nest and head for the sea at night to escape some of their enemies, but many are caught by ghost crabs.

enough light to find their prey but not so much that the prey can easily spot the appearance of the predators.

Among the preyed-upon fishes, the differentiation into nocturnal and diurnal is largely determined by their diet. Thus, the diurnal fishes are grazers, including the surgeonfishes which scrape algae from rocks, the parrotfishes which crunch lumps of coral, and the butterflyfishes which pick tiny animals from the surface of rocks or coral. The nocturnal species, such as squirrelfishes and wrasses, live by seizing shrimps and other invertebrates. It seems that diurnal fishes, the grazers, are dependent on light since they continue to feed on moonlit nights, but the activities of the nocturnal fishes are dictated by the habits of their prey which emerge from their hiding places among the rocks and seabed at night.

Still further inshore, on the margins of the sea, a new cycle of changes in the environment takes control of the lives of marine animals. This is the rhythm of the tides, which rise and fall in a sequence that varies in amplitude with the phases of the moon. It is of major importance because animals living within the tidal zone are periodically faced with the problems of virtually living on land at some time during the 24 hours, but they not only have a tidal rhythm, they also have a diurnal rhythm.

Thus, the crabs are, in general, active by night as a defence against predation. Added to attacks by carnivorous fishes and octopuses in the deeper waters beyond the shore, there are additional risks involved in living in the shallows and rock pools where waders, gulls and terns forage by day. So they react by being mainly nocturnal yet responding to the tidal rhythm. European shore crabs, for

153

Ghost crabs feeding on a dead fish at night.

example, come out from under stones or seaweed or from lying buried in the sand and forage for anything they can seize, live or dead and long-legged spider crabs climb fronds or seaweed to catch passing shrimps, all in the dark. On the other hand, hermit crabs can be seen out during the day because they are protected by the heavy shells which they have appropriated.

When animals first crawled out of the sea and evolved the ability to survive on land, over 400 million years ago, there could easily have been a tendency to confine activity to the night hours to escape the drying effects of the sun. Initially at least, there were no predators to evade or competitors to avoid. Indeed, these were good reasons for leaving the sea. It is difficult to imagine the scene in such remote times and especially to envisage the relationships between animals which may have been sharing the same habitats. Nowadays the seashore attracts many predators around the clock but the birds are a major influence only during the day and definitely cause small animals to come out at night.

One of the most striking examples of the use of nocturnalism by an otherwise diurnal animal is seen in the breeding habits of the sea turtles. Female sea turtles lay their eggs on beaches above the tideline, a task they do at night. They heave their heavy bodies up the shore and lay their eggs in the sand. Digging a pit to receive the eggs and filling it in again is a lengthy business and the turtles do not regain the water until the early hours of the morning. An adult turtle is a large animal and well armoured, so that it has little to fear on land, except man, but several hours physical exertion are best reserved for a time when the tropical sun is not beating on its shell.

154

Above: Having laid her eggs, a female green turtle heads for the sea as dawn breaks.

Grunions flop out of the waves to mate and lay their eggs at high tide.

In due course the baby turtles hatch out and make for the sea. They also use nocturnalism. They dig their way upwards, and pause just beneath the surface until a drop in temperature tells them that night has fallen. Then they break out and scamper helter-skelter down the beach, guided by the pale loom over the water. Baby turtles will suffer even more from hot sun and sand than their mothers, but the problem of predation is for them even more serious. Jackals, mongooses, monitor lizards and crabs gather to feast on them by both day or night, but the hatchlings will at least escape some of their enemies, such as gulls and other predatory birds, by moving at night.

The turtles' habit of restricting their vulnerable movements on land to the night is eminently sensible and they would make a fitting conclusion to a discourse on nocturnal behaviour. This would, however, not reflect the feeling of uncertainty which has pervaded the writing of this book. There are so few instances where it is possible to point to a clearcut reason for nocturnal behaviour or to clinch an argument with observations or experiments which demonstrate the advantage accruing to the animal. All too often, there are exceptions where the animal, or its close relative, is found to be active by day when by all reckonings it ought to be active by night. If there was evidence for a difference in behaviour or physique which accounted for the discrepancy, we could write QED and be pleased to find another example of the endless ways in which animal life has been moulded to fit the environment. Unfortunately, we cannot do this, as one final example shows.

Of the many tourist attractions in California, the most peculiar is the breeding of the grunion. This is a silvery fish, about 15 centimetres long, related to the sand smelts and silversides. It breeds at the new and full moons, when spring tides cover the beaches to the highest level. For three or four nights swarms of grunions appear in the surf and for an hour or so, as the tide is beginning to ebb, they flap and leap out of the water and onto the damp sand until the beach seethes with silver, writhing fish. Each female grunion wriggles into the sand to lay her eggs, while males gather around her to fertilise them. The process takes only a few minutes and the fishes quickly return to the sea. The eggs take a week to develop, but they do not hatch until they are shaken by the breaking waves of the next high spring tides so the hatchling grunions can launch themselves straight into the sea.

Hundreds of kilometres of coastline of the State of California and neighbouring Baja California witness the excursions of the grunions. Their spawning rituals are timed for the most suitable state of tide and the most suitable time of day. The darkness gives them immunity from attack by birds, which could not fail to be attracted by such an easily-won banquet. No one could dispute the value of this timing if it were not for the fact that another species of grunion, living around the corner in the Gulf of California, spawns by day and suffers from attacks by birds. Why should the two species behave differently? Is there an advantage for the first grunion in spawning by night, or is the timing dictated by another factor in its life? We have much to learn yet about nature's night life.

156

Credits

Index